Managing Livestock Wastes to Preserve Environmental Quality

Managing Livestock Wastes to Preserve Environmental Quality

J. Ronald Miner
Frank J. Humenik
Michael R. Overcash

Iowa State University Press
Ames

J. Ronald Miner, Professor and Extension Water Quality Specialist in the Bioresource Engineering Department, Oregon State University, holds a bachelor's degree in chemical engineering from the University of Kansas, a master's degree in civil (sanitary) engineering from the University of Michigan, and a doctorate in chemical engineering and microbiology from Kansas State University. He has been actively involved in livestock waste management teaching, research, and extension for the past 30 years.

Frank J. Humenik, Professor and Coordinator of Animal Waste Management Program, College of Agriculture and Life Sciences, North Carolina State University, holds a bachelor's degree in civil engineering and a master's and doctorate in civil (sanitary) engineering from Ohio State University. He has conducted research, teaching, and extension activities in waste management and water quality for the past 31 years.

Michael R. Overcash, Professor of Chemical Engineering, North Carolina State University, holds a bachelor's degree from North Carolina State University, a master's degree from the University of the New South Wales (as a Fulbright Scholar), and a doctorate from the University of Minnesota. He has been a leader in beneficial reuse of animal waste and the study of environmental effects of agricultural practices.

Iowa State University Press
Ames, Iowa 50014
Orders: 1-800-862-6657

Office: 1-515-292-0140
Fax: 1-515-292-3348
Web site: www.isupress.edu

⊗ Printed on acid-free paper in the United States of America

First edition, 2000

Library of Congress Cataloging-in-Publication Data

Miner, J. Ronald.
    Managing livestock wastes to preserve environmental quality / J. Ronald Miner, Frank J. Humenik, Michael R. Overcash.—1st ed.
        p. cm.
    Includes index.
    ISBN-0-8138-2635-7 (alk. paper)
        1. Animal waste—Environmental aspects. 2. Animal waste—Management. I. Humenik, Frank J. II. Overcash, Michael R. III. Title.

TD930.2.M55 2001
636.08'38—dc21   00-033440

Last digit is the print number: 9 8 7 6 5 4 3 2 1

Dedication

To Betty Sue Mary Rachel

Contents

Managing Livestock Wastes to Preserve Environmental Quality

1

Introduction

Livestock and poultry production has undergone an extensive transition since 1950. The number of farms producing livestock and poultry for the commercial market has decreased by more than 80%. Urban expansion has continued and consumers have maintained an interest in a diet rich in animal protein. The net result has been a dramatic change in the technology used by meat and milk animal producers.

Traditional animal agriculture includes a picture of an American farm with a diversity of animals and a flock of chickens being raised around the other labor demands of the crop production enterprise. It was not uncommon for the farm wife to be responsible for management of the chicken flock and market eggs and fryers as an ongoing source of available cash from more urban dwellers who came to the farm to purchase fresh products. The 1940s saw the development of confined poultry production in the United States based on housing the chickens in mechanically ventilated poultry buildings with mechanized feed distribution systems and automatic watering devices.

Other species were still being raised in small operations in 1950. Milk was being produced and sold by farmers with fewer than 20 cows. Milk was placed in cans and hauled to a buying station where it was accumulated for processing and eventual resale. Similarly, cattle were being produced on individual farms and taken to stockyards where competing meat processing organizations bid on the animals.

Although there is a romantic appeal to this picture of traditional livestock and poultry production, it is not the relevant picture as the 21st century begins. A variety of social, economic, and technological changes have taken place that have dramatically altered the way meat, milk, and eggs are produced and made available to the modern consumer. Among the changes were the following:

- There was a major exodus from the rural areas to growing industrial and commercial centers in response to regular assured incomes without the risk inherent in agricultural production. Many of these new

3

jobs had the additional benefits of regular hours, better pay, fringe benefits, and vacations.

- Researchers and innovative farmers demonstrated the benefits of mechanization in animal performance and labor efficiency. As labor costs increased, it became more desirable to automate processes that had previously required extensive manual input. Carrying feed and water to individual pens of animals around the farmstead was easily replaced by automated equipment.
- Growing nonfarm populations were affluent and created a demand for large quantities of high-quality meat, milk, and eggs. This demand was met by increasing herd sizes that incorporated mechanization. The taste for high-quality beef finished on high-energy feeds prompted the movement of cattle from pasture operations to confinement production where they could be fed a carefully controlled high-energy ration.
- Mass communication and advertising created demands for highly processed and convenient food products. Social changes were underway as well. The development of large-scale animal production in the Soviet Union on governmentally organized farms demonstrated the feasibility of what was then regarded as massive pig farms with highly specialized barns for particular stages in the life cycle.

Traditional livestock and poultry production had a by-product, manure, that was recognized as a valuable material for application to cropland. It was sufficiently valuable that it could be manually removed from stables with pitchforks and loaded into manure spreaders for hauling to cropland and applied to enhance production. This manure use was common before the availability of low-cost chemical fertilizers that were less burdensome to apply and less uncertain in nutrient content.

The growth of intensive livestock production began as individual innovative farmers started to construct facilities that would allow a single operator to manage more animals without the need to hire high-cost laborers. As with most other major technological changes, however, there were surprises along the way that punctuated the history of modern livestock and poultry production and that, in turn, created a need for livestock waste management.

The state of livestock production in the United States is reflected in Tables 1.1 and 1.2. As Table 1.1 shows, a few states dominate pig production. The number of farms decreases each year, whereas the number of pigs remains relatively constant. The top 10 pig producing states have slightly over half the total number of pig farms in the country, but produce over 80% of the pork. Table 1.2 reminds us that on a national basis, the largest 10% of the pig farms produces more than three-fourths of the market pigs. North Carolina, the state most prominent in the

Table 1.1. Leading U.S. hog-producing states in 1998 (U.S. Department of Agriculture, 1999)

State	Farms	Inventory (thousands)	% of Total U.S. inventory
Iowa	17,500	15,500	24
Illinois	7,000	4,850	8
North Carolina	4,200	10,300	16
Indiana	4,400	4,100	6
Minnesota	8,500	6,000	9
Nebraska	6,000	3,400	5
Missouri	5,000	3,350	5
Ohio	6,500	1,750	3
South Dakota	3,200	1,300	2
Kansas	2,600	1,550	2

recent move toward large-scale intensive pig production, has 98% of the pigs on farms of 1,000 or more animals. The smaller two-thirds of the farms produce only 2% of the market pigs. This trend toward intensification and specialization provides both challenges and opportunities.

Fish kills below Cattle Feedlots

During the early 1960s, it was considered good cattle feedlot design to build feedlots along steeply sloping, south-facing hillsides. By locating the feedbunks near the top of the hill and by having the pens extending downward, the access road would stay relatively dry even in rainy weather, which made feeding easier. The southern exposure also promoted rapid drying of the lot surface. Muddy conditions were

Table 1.2. Hog farm size distribution within states, 1998 (U.S. Department of Agriculture, 1999)

	Farms with <1,000 head		Farms with >1,000 head	
State	Farms	Inventory (% of state total)	Farms	Inventory (% of state total)
Iowa	13,300	25	4,200	75
Illinois	5,750	27	1,250	73
Indiana	5,340	27	1,060	73
Minnesota	7,100	26	1,400	74
Nebraska	5,280	38	720	62
Missouri	4,470	23	530	77
North Carolina	2,730	2	1,470	98
Ohio	6,050	45	450	55
South Dakota	2,900	37	300	63
Kansas	2,390	21	210	79
Total U.S.	55,310	23	11,590	77

recognized as detrimental to cattle performance. This design had the additional benefit of having the manure worked down the hillside by cattle activity.

Although fish kills were not unknown in the 1960s, fish kills in agricultural areas were uncommon. There had been reports, however, of fish dying in unexpected rural locations, and even more surprising, during and immediately after rainfall when dilution should have been available. In response to one of these reported atypical fish kill events, samples were collected downstream of a feedlot. When the data were analyzed, a standard but moving oxygen sag curve was detected. Figure 1.1 shows how runoff from the feedlot first entered the stream, began to support oxygen-consuming bacteria, depleted the available oxygen, and caused the fish to die.

This incident along with others similar but possibly less dramatic led to the observation that runoff from livestock feedlots was a concentrated source of water pollutants capable of causing extensive and widespread water pollution. Initial reactions were that the runoff was beyond control because it represented an act of nature. Second thoughts, however, prevailed and over the next decade the design and construction of cattle feedlot runoff control structures became standardized and cattle feedlots equipped with these structures were commonplace. Today, cattle feedlot design includes runoff control provisions that protect nearby waterways.

Swine Confinement Building Smell Plagues Neighbor

An affluent investor in the Midwest constructed a swine confinement building on an abandoned farmstead because of the availability of an existing feed storage facility. This early confinement structure was complete with underfloor manure storage, an automated water distribution system, and a feed-handling system that reduced the routine labor requirement to tend the 450 finishing pigs to less than 1 hour per day. Although the investor had no previous experience with pig production, he or she was able to handle the routine labor without difficulty in addition to other activities.

Unfortunately, the location selected for the confinement building was less than 150 meters southwest of the home of a long-established elderly farm family who raised a variety of crops and fed a small number of pigs in dirt lots immediately north of their home. On this farm, the owner carried both feed and water to 15 pigs twice daily in a highly labor-intensive system much as was done for more than 40 years.

The confinement building clearly had an odor. This odor proved to be so severe it caused the elderly neighbors to seek medical care for

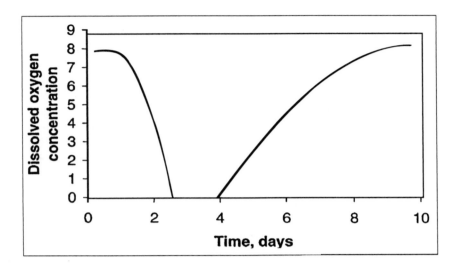

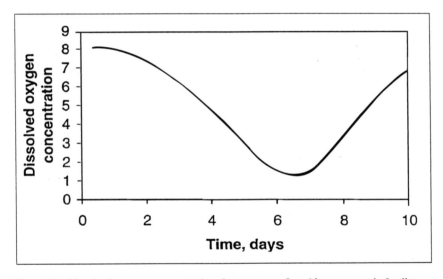

Figure 1.1. Dissolved oxygen concentration downstream of a midwestern cattle feedlot, 1962. Notice that during the first sampling (top), the runoff had traveled a relatively short distance downstream of the entry point but was sufficiently concentrated to have totally depleted the available oxygen. The conditions 4 days later (bottom) when the runoff had moved farther downstream were considerably dispersed, impacted a longer reach of the stream, and had lowered dissolved oxygen below that required for survival of most game fish.

their problems and to flee their home on numerous occasions for a less odorous location. This new odor eventually caused them to seek a court order against the facility as a legal nuisance. Testimony in the case established that both the confinement building and the more traditional mode of pig production were sources of odor. The impacted neighbors were not bothered by the smell arising from their operation but were indeed impacted by this new nontraditional odor.

In retrospect, this example highlights the complexity of odor control associated with livestock and poultry production. Clearly, the new facility was located too close to an established residence and it lacked odor minimization design features that are standard in contemporary designs. More importantly, this experience raises the issues of subjective response associated with manure odors and the extent to which odors trigger emotional responses in people that are much more severe than an objective evaluation would predict. The odor of the confinement barn reminded the neighbors of the injustice associated with their labor-intensive system and the alternative highly capital-intensive but low-labor alternative. A court order that the confinement building be moved to an alternate location provided partial relief for the impacted neighbors but proved an expensive solution for the investor.

As students of livestock waste management, this episode reminds us of the importance of site selection and facility design. In addition, it helps us understand the emotional response of neighbors to odors.

Perspective

These two examples were selected to highlight the real world of livestock waste management in a contemporary world. There are many resources available to guide the design, construction, and operation of livestock facilities, yet it is a dynamic endeavor that is continually changing in response to new information, new regulations, and an ever-changing society with growing expectations for an environmentally benign food production scheme. This book was designed to provide a compilation of the experiences of livestock waste management engineers and scientists gathered over the past 40 years in the hope that it may spark new ideas on how to more adequately meet the challenges of a safe, affordable, high-quality food supply that respects the environment in which we work, play, and imagine new and exciting alternatives. More specifically, the purposes of the book are as follows:

• To provide the environmentally concerned professional with an understanding of the challenges facing the livestock producer in managing animal wastes in an environmentally responsible manner.

- To provide the livestock producer with an understanding of the potential impact of mismanaged livestock wastes and of the importance of rigorous design, careful construction, responsible management, and devoted maintenance if a manure management system is to function effectively.
- To provide the technically inclined student or practitioner with the basic principles and tools necessary for the conceptualization of a livestock waste management plan that can be tailored to local conditions, needs, and capabilities to ensure sustainable production that protects soil, water, and air resources.

2

Potential Impact of Unmanaged Livestock Waste on the Environment

Livestock waste generally includes manure and urine; however, it is appropriate in many situations to consider a broader range of materials and their impact on the environment. For example, the disposal of dead animals is an important consideration whether in small or large operations. Waste materials also might include spoiled feed ingredients, silage drainage, spent dipping vat chemicals, or other materials only partially consumed in a livestock or poultry operation.

Environmental pollution generally involves damage to a shared resource, such as soil, water, or air. Both surface and underground water need to be considered. Damage to soils also may be of concern in certain situations; however, because the soil is commonly fixed in space, pollution takes on a more restricted meaning.

Livestock and Poultry Manure: Nature and Quantity

Livestock manure includes both the feces and urine of the animals involved. Extensive data exist characterizing the quantity and chemical characteristics of manure. Table 2.1 summarizes these data in sufficient detail for most uses. It should be evident that these data represent the collective judgement of several researchers and are typical numbers that may not represent what would be found in any particular circumstance. There are individual differences in manure quantity among animals and between breeds, and quantity also can be effected by whether an animal is being fed to gain weight or produce milk or is fed on a limited-maintenance ration. The important point is that data are available to predict manure quantities as excreted. The challenge, however, is that the housing and manure collection system has a major impact on the manure characteristics to be considered in any management system.

Table 2.1 shows that fresh animal manure is more than 75% but less than 95% moisture. Thus, it is neither a liquid nor a sufficiently dry solid to allow handling as a typical solid. Manure handling systems are there-

10

Table 2.1. Manure quantities from various animals

Animal	Animal size (lb)	Manure produced (lb/day)	Manure produced (ft³/day)	Moisture content (%)
Dairy cow	500	41	0.66	87
	1,000	82	1.32	"
	1,400	115	1.85	"
Beef cattle	500	30	0.50	88
	1,000	60	1.00	"
Swine				
Nursery pig	35	2.3	0.04	91
Growing pig	65	4.2	0.07	"
Finishing pig	150	9.8	0.16	"
Gestating sow	275	8.9	0.15	"
Sheep	100	4.0	0.062	75
Poultry				
Layers	4	0.21	0.0035	75
Broilers	2	0.14	0.0024	"
Horse	1,000	45	0.75	80

fore designed to either add sufficient water to allow handling as a liquid or to promote sufficient drying to allow handling as a solid. Liquid manure systems typically require a water addition, and they operate best if the amount of water added is restricted to that necessary for effective functioning of the system. Excess water increases treatment and transport costs with no benefit to the operator beyond transport from the building. One of the functions of manure handling systems is to adjust the physical form to facilitate handling for ultimate disposal. Traditionally, manure was converted to a dry material by the use of bedding so it could be hauled to cropland or pastures with a manure spreader. Liquid systems have become more popular in recent decades so the final application could be accomplished with specially adapted irrigation equipment.

Manure is not a solution but is a suspension of particles in a liquid rich in dissolved organic material. This understanding is important as we begin to consider management options. Many of the current management systems include solid–liquid separation to facilitate further handling, byproduct recovery, or treatment.

Potential Impact of Organic Matter in Manure

Manure is rich in organic material. A portion of this organic material is undigested or partially digested feed ingredients. Careful examination of manure reveals particles that can be identified from the feed ingredients. The complex organic compounds of livestock rations are degraded into simpler molecules by biochemical processes during energy production. For example, carbohydrates are converted into starches, then into sugars, then into organic acids, alcohols, or ketones, and finally into carbon dioxide, methane, and water. Proteins similarly are broken down into amino acids, ammonia, urea, sulfides, and other simpler materials. This aspect of animal physiology explains why manure is not easily characterized by conventional chemical analyses and why the number of odorous compounds in manure is in the hundreds and potentially even greater quantities, as detected by ever more sensitive analytic techniques.

Livestock and poultry manure, like other organic wastes, is available for further bacterial decomposition once it enters the environment. If the manure enters water that contains dissolved oxygen such as a natural stream or impoundment, an ideal situation is created for resident aerobic bacteria to thrive on the manure components as a food supply. This process is similar to that which occurs if human or domestic sewage enters a watercourse. Aerobic bacteria use oxygen as the essential ingredient in their metabolic energy production process. Therefore, if manure enters a body of water in which dissolved oxygen is available and the temperature is appropriate, the bacteria thrive, converting the manure components into simpler compounds and, in the process, they consume the available dissolved oxygen. The extent to which bacteria consume the available oxygen depends on the amount of manure available and the number of bacteria present. Because the bacteria involved have relatively short generation times, the bacterial population is not a constraint in metabolizing waste nutrients and in depleting the limited amount of oxygen in surface waters.

Dissolved Oxygen in Natural Waters

Oxygen is an essential for humans. Similarly, it is critical for aquatic organisms and is among the factors that determine the nature of the population that resides in a watercourse or impoundment. Thus, as long as fish are considered a critically important part of a quality watercourse, stream, river, or lake, the dissolved oxygen content of the water is a critical parameter. Were dissolved oxygen not such a critical feature of water quality, it might be judged insoluble by conventional standards. It requires 15,000 gallons of water at saturation to provide a single pound of dissolved oxygen. Table 2.2 contains values for the saturated, dissolved oxygen content of water at various temperatures, at sea level, and under a standard one atmosphere of pressure. Obviously, the saturation values are dependent on temperature, pressure, and the presence of other materials in the water.

Table 2.2. Saturated dissolved oxygen concentrations in water at various temperatures

Temperature (°C)	Dissolved oxygen concentration (mg/l)
0	14.6
10	11.3
20	9.2
30	7.6

In attempting to characterize the impact of organic materials on water quality, environmental scientists have created the concept of biochemical oxygen demand (BOD). BOD is an indirect measure of the biologically available organic matter in a natural water or in a waste material based on the amount of oxygen an abundant aerobic bacterial population would consume in metabolizing that material as an energy source. Conceptually, it is as though the following reaction is underway:

$$BOD + O_2 \rightarrow CO_2 + H_2O + Energy \tag{2.1}$$

The model further assumes that food (BOD) is consumed in proportion to the amount present at any particular time, i.e., a first-order reaction, which is mathematically stated as follows:

$$\frac{-dL}{dt} = k'L \tag{2.2}$$

where L = the amount of biologically oxidizable organic matter available, or alternatively the amount of oxygen that would ultimately be consumed, and k' = a rate constant representing the fraction of the organic matter consumed per day. This statement of the concept can be integrated to yield a more useful expression:

$$L = e^{-k't} = 10^{-kt} \tag{2.3}$$

where $k = k'/2.303$.

Equation 2.3 is the classical expression of first-stage BOD, which has been used for decades to describe the organic content of waste materials in the most useful fashion, namely, the amount of oxygen that is consumed by aerobic bacteria in stabilization. Frequently, this expression is more useful if it is rearranged to state the amount of BOD exerted in a particular time:

$$y = L (1 - 10^{-kt}) \tag{2.4}$$

where y = BOD exerted at any time, t, and L = the amount of ultimate BOD.

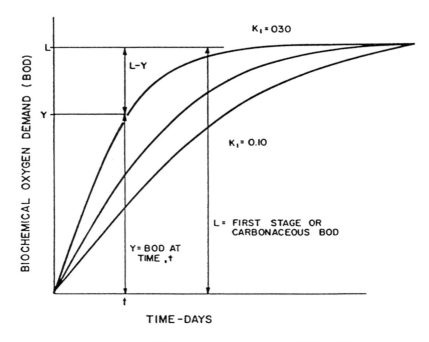

Figure 2.1. The progression of biochemical oxygen demand with time. These drawings show the progression at three different rate constants. Higher rates would indicate a greater degree of biological availability. Lower rates suggest greater resistance to aerobic bacterial degradation. Rate constants also are influenced by temperature.

Equations 2.3 and 2.4 are classical parabolic expressions, which can be plotted as shown in Figure 2.1. The expressions developed describe what has traditionally been called carbonaceous or first-stage BOD. Usually, when practitioners refer to BOD, they are referring to the material and the model. If BOD analyses are conducted on a naturally occurring organic waste, including livestock waste, the data follow this line for several days and then there is normally an additional rise not predicted by this curve that is known as second-stage or nitrogenous BOD. This additional oxygen consumption is due to the nitrification of reduced nitrogen compounds that are included in the waste.

Often, BOD is discussed in the literature as the 5-day BOD or BOD_5 because of the need to complete the analysis in a reasonable period of time and the desire to get a measured value before the onset of second-stage BOD, nitrification.

The value of k, the rate constant in equation 2.4, can be determined experimentally by conducting a series of BOD analyses over time and plotting the data as shown in Figure 2.1. Typically, the value of k is taken to be 0.1 per day at 20°C unless more specific data are available. The value of k at 20°C reflects the biodegradability of the wastewater to the mix of aerobic bacteria that were used in conducting the analysis. Techniques for conducting the BOD analysis are well defined (APHA, 1995).

As would be expected, the value of k is temperature dependent. Values determined at 20°C can be adjusted with equation 2.5:

$$k_t = (k_{20})1.047^{(t-20)} \qquad (2.5)$$

The coefficient of 1.047 in equation 2.5 is typically used. More specific values can be determined experimentally if needed.

Oxygen Balance in Natural Waters

The impact of an organic waste discharged into a stream or lake can be predicted by considering the BOD expression just developed and a similar expression for natural reaeration. Reaeration is the process in which oxygen dissolves in water through the air–water interface. The greater the turbulence the greater the amount of interchange. Stream reaeration is important as a self-purification process allowing a waste-assimilative capacity to natural waters. The rate of reaeration is considered to be proportional to the dissolved oxygen deficit at a particular time:

$$\frac{dC}{dt} = k_2 (C_S - C) \qquad (2.6)$$

where C = dissolved oxygen concentration at any time, milligrams per liter; C_S = saturation dissolved oxygen concentration, milligrams per liter; and k_2 = reaeration constant, days.

A similar equation develops if the term $C_S - C$ is replaced by the term D, the dissolved oxygen deficit:

$$\frac{dD}{dt} = -k_2 D. \qquad (2.7)$$

This expression can be integrated as was done for equation 2.2 to give

$$D = D_0 10^{-k_2' t} \qquad (2.8)$$

where D = dissolved oxygen deficit at any time, D_0 = initial dissolved oxygen deficit, and k_2' = reaeration constant, base 10.

The equation for the oxygen removal from water (2.3), and the reaeration equation (2.7) can be combined and integrated (Overcash et al., 1983) to yield the classic oxygen sag equation that relates the initial dissolved oxygen in a stream, the BOD at the waste entry point, and downstream dissolved oxygen deficits:

$$D = \frac{k_1 L}{k_2 - k_1} (10^{-k_1 t} - 10^{-k_2 t}) + Da\, 10^{-k_2 t} \qquad (2.9)$$

where k_1 = BOD rate constant, base 10; and k = reaeration rate constant, base 10.

Two other important items, the time of maximum deficit and the magnitude of that deficit, also can be calculated.

$$t_c = \frac{1}{k_1 (f - 1)} \quad 2 \cdot 3 \log \left\{ f \left[1 - (f - 1) \frac{D_A}{L_A} \right] \right\} \qquad (2.10)$$

$$D_c = L\, 10^{-k_1 t_c / f}$$

where: $f = k/k$ (self-purification constant).

Thus, by using this analysis it is possible to calculate the impact of a discharged organic waste on the dissolved oxygen content of a receiving stream as a function of the initial dissolved oxygen and the characteristics of the input. Figure 2.2 shows this relationship.

What is important in this analysis is the observation that the organic matter contained in livestock and poultry waste can be treated similarly to other waste materials in predicting the impact on dissolved oxygen concentration of surface waters. These waste streams typically have BOD concentrations in excess of 5,000 milligrams per liter compared with approximately 200 milligrams per liter for municipal wastewater. This high strength generally precludes livestock waste from being discharged into receiving streams. The BOD production of various livestock and poultry are shown in Table 2.3.

Nutrient Content of Animal Manure

Livestock and poultry wastes have historically been recycled to the environment as a plant nutrient to support crop production. Regardless of the scheme to raise animals, there was a provision to incorporate manure back onto land to claim the economic benefit. In addition to the economic benefit of manure application to cropland, there is a major environmental benefit in having nutrients, especially nitrogen and phosphorus, on land rather than in receiving streams. Liquid manure tends to have the majority of its nitrogen in the ammonia form, which is not only a plant nutrient but also toxic to the aquatic environment. Table 2.4 lists the nutrient content of animal manure on a daily basis as excreted. In

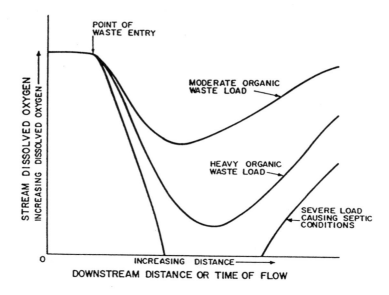

Figure 2.2.   The oxygen sag equation allows for prediction of the dissolved oxygen concentration downstream of a waste discharge. This approach can be used to estimate the extent to which a wastewater must be treated to permit its discharge without adverse impact on downstream water users.

Table 2.3.   BOD in the daily manure production of various animals

Animal	Animal weight (lb)	BOD produced (lb/day)
Dairy cow	500	0.86
	1,000	1.70
	1,400	2.38
Beef cattle	1,000	1.60
	1,250	2.00
Swine		
Growing pig	65	0.20
Growing pig	200	0.63
Sow and litter	375	0.68
Poultry		
Layers	4	0.014
Broilers	2	0.0023

Table 2.4. Number of animals producing sufficient manure to fertilize 1 acre of cropland based on 200 lb of nitrogen, 40 lb of $P_2 O_5$, and 30 lb of K_2O (assuming no loss of nutrients)

Animal	Animal size (lb)	Based on nitrogen	Based on phosphorus ($P_2 O_5$)	Based on potassium (K_2O)
Dairy cow	500	3	2	0.5
	1,000	2	1	0.3
	1,400	1	0.3	0.2
Beef cattle	500	3	1	0.5
	1,000	1.6	0.5	0.3
Swine				
Nursery pig	35	35	9	7
Growing pig	65	18	5	3.5
Finishing pig	150	8	2	1.5
Gestating sow	275	9	2	1
Sheep	100	12	7	2
Poultry				
Layers	4	190	43	55
Broilers	2	235	93	96
Horse	1,000	2	1	0.5

Source: Adapted from Midwest Plan Service, Livestock Waste Facilities Handbook, MWPS-18, 1985, Ames: Iowa State University, MWPS.

interpreting these values, be aware that for different housing and manure collection and manure management systems, a portion of the nutrients is lost. This is particularly true for nitrogen because several volatile forms exist. Phosphorus and potassium are less subject to volatilization.

Nitrogen Concerns

Nitrogen is of concern to water quality because of the solubility of its oxidized form, nitrate. Nitrate is not only a plant nutrient with implications in promoting excess plant growth in impounded waters but also presents a health concern in drinking water. Land application of manure at rates and times that exceed recommended agronomic procedures or crop uptake abilities results in nitrogen transport to groundwater and, frequently, surface water. Nitrogen in animal manure is present generally as both ammonia and organic nitrogen. In anaerobic storage, such as manure storage pits or anaerobic lagoons, organic nitrogen is continually being converted to ammonia and a portion of the ammonia is being volatilized and lost to the air, contributing to both odor and to areawide acid rain and nitrogen enrichment of surface water concerns. When manure is applied to cropland and incorporated into the soil as a plant nutrient, the environment changes from anaerobic to aerobic. Initially, the ammonia, present in part as ammonium, is attracted to the negatively charged soil particles and temporarily immobilized. Aerobic bacteria become involved and initiate the conversion of ammonia to nitrite then to nitrate. Generally, in agricultural soils the fraction of the nitrogen detected as nitrite is very small and can be ignored. Nitrate, however, can be present in large concentrations, up to several hundred milligrams per liter. As water moves downward through the soil, nitrate is carried with the water and nitrate concentrations in groundwater increase. In certain agricultural areas, the nitrate concentration in a significant percentage of the domestic wells exceeds the drinking water standard of 10 milligrams per liter nitrate as nitrogen.

Nitrate in drinking water is of particular concern because of its health effects in infants under 6 months of age as well as the hazard to pregnant and lactating women. The enzyme system of human infants converts nitrate to nitrite. Nitrite tends to bind with hemoglobin, thereby restricting the ability of the blood to transport oxygen. Although the oxygen–hemoglobin binding is loose, the nitrite–hemoglobin binding is more permanent. Thus, high nitrate concentrations are correlated with elevated rates of aborted pregnancies and infant deaths. An infant suffering from methemoglobinemia shows the blue color of persons lacking adequate oxygen in their blood. Because of the potential for development of this dangerous condition, it is appropriate for rural residents who are using individual wells to have their water supply tested for nitrate concentration on an annual basis. Although the connection between nitrate concentrations and animal productivity is less well understood, there is an overall thought that high levels of nitrate in the overall ration of animals contributes to more frequent abortions in cattle and more frequent death of calves. For cattle, there may be other sources of nitrate in addition to water supplies. Forages that have been highly fertilized and then suffer water deficiency tend to accumulate nitrates in sufficient concentrations to be of concern to animal health.

Nitrogen from livestock waste is generally regarded as a groundwater concern because of the infinite solubility of nitrate. Thus, in planning manure management and use systems, nutrient management is of paramount importance. This topic is treated in detail in Chapter 10 but it is appropriate herein to indicate the importance of the issue and to consider the long-term nature of the problem. Depending upon the type of subsurface soil layers, rainfall in an area, and the overlying soil characteristics, groundwater moves very slowly. Typically, groundwater is thought to move from less than one to a few meters per month. Thus, there is likely to be a long delay between the application of animal manure and any change in the nitrate concentration of nearby drinking water supplies. This factor complicates monitoring efforts to prevent groundwater pollution, and it greatly increases the risk associated with inadequate manure management and serves as a reminder that pollution prevention is far superior to remediation. If groundwater becomes contaminated, it is likely that several years will be required for the aquifer to recover.

Phosphorus from Manure as a Pollutant

Phosphorus is the second most plentiful nutrient in manure but has received less attention than nitrogen because of its limited solubility in water and its lack of toxicity to humans. Phosphorus is an essential nutrient to animals and is also essential to the growth of algae in water impoundments. As an essential element in algal growth, it is often identified as the control nutrient to prevent excess algae from developing.

In the past and even today, it is common to apply manure to cropland according to the nitrogen needs of the crop. Frequently, this means that phosphorus and potassium are applied well in excess of that being removed by the crop. Because potassium is water soluble, it tends to move through the soil profile and is not considered a problem. Phosphorus, however, is quickly bound to soil particles upon application and becomes part of the soil phosphorus reservoir. Whenever erosion occurs in areas that have received the excess phosphorus, the heavily phosphorus-laden soil particles serve as a major contribution to the phosphorus content of receiving lakes and reservoirs. This problem has been most notably identified in the intensive dairy areas of Florida in which runoff eventually reaches the Everglades. The response in this area has been to regulate the application of manure to cropland based on phosphorus content rather than on nitrogen content. The net impact of such a decision is to increase the land area required for manure disposal. Other areas of the United States are considering similar approaches because regulatory agencies are becoming increasingly concerned about phosphorus in natural waters and phosphorus buildup in soils. The oversupply of nutrients, especially nitrogen and phosphorus, to receiving waters that results in excessive vegetative growth and subsequent water quality degradation is defined as eutrophication.

Point Source Pollution Versus Nonpoint Source (NPS) Pollution

In simplest terms, point sources of pollution are ones that flow from a pipe, ditch, or other identifiable conveyance into a receiving water. NPSs of pollution may be considered as diffuse or coming from an area rather than through an identified conveyance. Large livestock operations, those with more than 1,000 head of beef cattle or an equivalent number of other animals, are defined as point sources for regulatory purposes. Small-animal enterprises or those that lack a defined waste discharge point are generally included in NPS pollution. Other sources of NPS pollution are runoff from cropland; stormwater runoff from other areas that flows naturally into streams and reservoirs; small, individual sewage treatment systems; and grazing livestock. In certain areas of the western United States, there is major concern over the impact of grazing livestock on water quality.

Grazing livestock are a classic example of NPS pollution because under many conditions, there is little if any identified impact. When numbers of animals increase and when these animals congregate in the riparian area of a stream, they defecate in the stream as they drink and tend to damage the streambanks as they graze and climb the banks.

In subsequent chapters, the regulatory process for controlling NPS pollution is considered in greater detail but herein it is sufficient to consider the matter and to recognize its importance in environmental quality. Recent assessments suggest that NPS pollution is associated with the majority of water quality-limited waters in the United States and that agriculturally related NPS pollution is the prominent source. This is an important challenge because a large portion of the land area is identified as agricultural land. Forestry is another major land use for which NPS pollution has become an important consideration.

Livestock Waste and Disease Transmission

Livestock and poultry wastes contain a variety of organisms associated with the intestinal tract of these animals and they are detected in the same manner as those of human origin. Fecal coliform bacteria are a specific group that can be identified and enumerated in both livestock and poultry waste. Indicator organisms are a more general and larger group of organisms that are easily counted and are therefore used to indicate the potential presence of pathogenic or disease-producing organisms. As a net result, there are concerns about the safety of waters into which animals have defecated or those into which livestock waste has entered either from point sources or where runoff has carried manure into water.

Although the frequency of human diseases due to livestock and poultry wastes is small, the potential exists and there are classic accounts of human contact with waters contaminated with animal manure that have resulted in disease outbreaks among humans. Diseases transmissible from livestock to humans are called zoonoses. Table 2.5 lists the more common zoonotic diseases. The most common response to the prevention of disease transmission from livestock and poultry to humans is to manage the wastes in response to the potential diseases that may be transmitted. Appropriate sanitation practices are suggested for persons who

Table 2.5.   Diseases transmissible from livestock to humans

Bacterial	Viral	Other
Salmonellosis	Anthrax	Q fever
Staphylococcus	Foot and mouth disease	Trichinosis
Tetanus	Ornithosis	Histoplasmosis
Tuberculosis		
Brucellosis		
Leptospirosis		
Colibacillosis		
Coccidiosis		
Tularemia		
Encephalitis		
Erysipelosis		
Infectious bronchitis		
Newcastle disease		

Source: Overcash, Michael R., Frank. J. Humenik, and J. Ronald Miner. 1983. Livestock
    Waste Management, CRC Press, Boca Raton, FL.

may come into contact with manure or water contaminated with manure. The first step is to avoid direct contact; the second is frequent hand washing, particularly before eating or drinking. Many people working with manure elect to wear plastic gloves, protective garments, and dust masks for protection of their personal health. Regular inoculations for typhoid are important preventatives as are regular tuberculosis tests.

Perhaps as important as the direct transmission of disease from livestock and poultry to humans is the management of livestock wastes to meet water- and air-quality standards according to the applicable standards of the area. For example, there are several locations where there are large concentrations of livestock immediately above a water use or withdraw point that requires particular management. Along the west coast of Oregon, there is a concentration of dairy farms in a watershed that drains into a bay from which oysters are harvested for human consumption. This region also happens to receive abundant wintertime rains from 2,000 to 4,000 millimeters.

The traditional practice in Oregon was to store manure in storage tanks of limited capacity and to apply manure to nearby pastureland throughout the year. It was noted that at times when the river stage rose after a storm, the fecal coliform (FC) concentration in the bay exceeded the allowable standards for harvesting oysters (14 FC/100 milliliters). The problem was diagnosed as being at least partially due to runoff from pasture areas into nearby streams that were tributary to the bay. The situation was particularly severe when the soils were saturated, when manure was applied during rainfall, or when rainfall occurred immediately after manure application. The solution was to provide additional manure storage capacity on the dairy farms so the operators had greater flexibil-

Table 2.6. Estimated production of indicator organisms by humans and selected domestic animals

Species	Density FC, millions/g	Density fecal streptococci, millions/g	Total FC/ animal/ day, billions	Total fecal streptococci/ animal/day, billions	Ratio, FC to fecal streptococci
Human	13	3.0	2.0	0.45	4.3
Chicken	1.3	3.4	0.24	0.62	0.4
Cow	0.23	1.3	5.4	30	0.2
Duck	33	54	11	18	0.6
Pig	3.3	84	8.9	23	0.04
Sheep	16	38	18	43	0.4

ity in when to apply manure. In this example and many others, the elevated FC concentrations were not exclusively related to dairy manure; there was also a contribution from malfunctioning local sewage treatment facilities and individual sewerage systems.

Importantly, the number of organisms that indicate the potential presence of pathogenic organisms can be reduced by application of improved waste management facilities; however, daily management of these facilities is critical to the success of the improved systems. For example, if a dairy farmer builds manure storage capacity for 90 days and allows the tank to fill even though appropriate days for manure spreading had occurred, the benefit of the storage may be lost. A full manure storage tank does not provide any flexibility in when to apply manure. In this example, the solution is to make manure management a priority in how the operator schedules his or her time and effort.

Livestock and poultry, like humans, produce great numbers of indicator organisms. Table 2.6 summarizes some order-of-magnitude data that are helpful in understanding this situation. The most obvious implication of these numbers is that manure is such a rich source of indicator organisms that any treatment system that is going to produce an effluent for discharge needs to have an indicator organism removal efficiency well in excess of 99.9%. Manure management systems in the United States have not been designed for stream discharge but have been designed to facilitate alternate use and disposal methods. The predominate site for manure disposal has been crop and pastureland.

Livestock and Poultry Waste Discharges into Air

Airborne discharges include dust and other particulate matter, reduced gases from anaerobic decomposition, and odors. Odors are the greatest challenge of the three airborne discharges.

Dust issues arise when livestock are confined in feedlots in the arid portions of the country. Cattle feedlots have received the greatest amount of attention because of the widespread confinement of cattle in feedlots during the final months of their production to achieve a high-quality product for the consumer. In these feedlots, the vegetation is completely removed and a surface manure pack develops. The action of the animal hooves on this manure pack creates huge clouds of dust if some control is not exercised. Two approaches to this problem are developed later in the text. Briefly, the first is to increase animal density in feedlots during the dry period of the year and to use the additional urine that is deposited as an assist in maintaining the lot surface sufficiently moist to reduce dust production. Lower densities are typically maintained during periods of the year when it is cooler, when there is more frequent rainfall, or when muddy conditions are anticipated. Feedlot management also may include lot shaping to promote improved footing for the animals. Sprinkling with water is the second management option. Sprinkling systems to control dust in feedlots require careful design to apply water at the appropriate rate to control dust but not at a rate sufficient to create muddy conditions or wet areas that would promote anaerobic decomposition with the associated elevated odors.

Anaerobic decomposition of manure is a complex biological process with a large number of intermediate products. Upon exposure to air, a fraction of these compounds escapes and becomes airborne. Methane is one of the final products of anaerobic decomposition and is evolved from many treatment processes. Methane is also one of the "greenhouse gases" thought to be contributing to global warming. Although no regulatory constraints currently exist concerning methane emissions from livestock production facilities, concerns have been expressed and they are not likely to ease in the future. Ammonia also is evolved from a number of manure treatment and storage processes. Ammonia in the air is of concern for three reasons: 1) it is deposited on nearby water surfaces, 2) it is part of acid rain, and 3) it is related to odors arising from livestock production. Furthermore, ammonia is one of the more easily detected and measured odor-related gases. Hydrogen sulfide also is released during anaerobic manure decomposition and is of concern both because of its toxicity and its involvement in malodor.

Odor, the last of the three airborne pollutants, is currently the most serious threat to livestock producers in the United States and many other countries. Odors are a human phenomenon of great social importance. The smell of nearby manure decomposition is unacceptable to a sufficiently large portion of the population and has created concerns throughout the livestock and poultry industry. Techniques exist to design and operate livestock waste management facilities that produce fewer odors, however, they are generally more expensive to construct and operate than the more odorous counterparts. Because odor is a localized concern, the traditional approach has been to select a site sufficiently remote from neighboring residences so that odor control technology is not required. With livestock and poultry operations growing in size and the rapid development of rural areas, identifying sites with adequate separation has become increasingly difficult. The public also has become less tolerant of these odors, in part, because the operations are owned by "others" rather than "neighbors" and all the benefits of ownership return to these "others" rather than to members of the local community.

Reference

American Public Health Association. 1995. Standard Methods for the Examination of Water and Wastewater, 18th Ed. Washington, D.C.

3

Public Response to Environmental Quality Impact of Livestock Production

Whenever U.S. citizens decide there is a problem, that perception is generally reflected in the actions of their elected representatives. As a result, problems usually get attention when there is an impact on a sufficiently large population or a sufficiently strong impact on a limited constituency to prompt legislative action. Water quality legislation arose in the U.S. congress before the turn of the century, when in 1897 the Rivers and Harbors Act was passed to prohibit the dumping of solid wastes into any navigable waters within the boundaries of the nation. This initial activity was primarily devoted to the protection of shipping and water transport but began to set the pattern that waters were indeed public property and were not available for an individual or a particular interest group to so dominate a waterway that it became unusable to others.

Water Quality Concerns

At the time water quality regulation was being considered, the concept of individual state's rights was important. Therefore, the next round of regulatory activity was within individual states as they began to discover that the uncontrolled discharge of wastes from sewers, and particularly from manufacturing and food processing industries, began to create water quality problems. Early in the 20th century, most people lived in small communities or in rural areas, and there was limited running water and hence little demand for sewer systems or wastewater management. Livestock and poultry operations were small and related to individual farms, therefore, there was no livestock waste handling problem.

As the individual states continued to manage the water quality in their jurisdictions, an obvious problem arose. What do states do when the wastewater discharges in one state flow across the state line and create problems in the downstream state? The downstream state had no jurisdiction in the state where the problem was being created and the upstream state was not suffering a problem. This situation was developing along the Ohio River in Pennsylvania, Ohio, Indiana, and Kentucky.

26

Federal Legislation

The Water Pollution Control Act (PL 80-845), passed in 1948, was a landmark in water quality regulation because it established that there was federal responsibility to control pollution of interstate waters. It also made an initial allocation of federal funds to share in the cost of constructing water pollution control facilities. Livestock waste was not an issue at this point, however, discharges from livestock and poultry slaughtering plants were issues and some of the early efforts were devoted to cleaning these operations. At this point in time, water quality concerns were generally not of an environmental quality nature but were more in response to the public health concerns of disease transmission.

In 1935, the link between agriculture and water quality was recognized by the Soil Conservation Service. It involved rural landowners in planning and adopting practices that reduced erosion of cropland into streams.

By 1965, however, the public had discovered the concept of environmental quality and Congress passed water quality legislation that was to shape the future of waste management. The 1965 Water Quality Act established the Federal Water Pollution Control Administration, which would eventually become the Environmental Pollution Control Agency (EPA). It directed each state to establish highest and best-use designation for its waters, to establish water quality standards appropriate for that use, and to develop a compliance plan to achieve and maintain these standards. For most streams in the United States, this meant that water quality was to be maintained at a level necessary for the native fish population.

National Environmental Policy Act. Around this same time, feedlots were beginning to develop for the final feeding phase in beef production, the first fish kills due to feedlot pollution were noted, and there was an increasing awareness that livestock waste management might be necessary. Individual states were beginning to give thought to these issues, and municipal wastewater treatment plants were being built around the country. Communities that had previously been served by individual septic tank systems and outhouses were building sewage collection systems and lagoons, and waste stabilization ponds were being constructed at a rapid rate with federal cost share programs and federal oversight. The National Environmental Policy Act of 1969 required that all federal agencies explicitly consider environmental impacts in reviewing proposed programs. Responsibility at the federal level for water pollution control was shifted from the Department of Health, Education, and Welfare to the Department of the Interior, and finally to the newly established

Environmental Protection Agency. An environmental awareness was developing throughout the country that surpassed any that had previously been known.

Public Law 92-500, National Pollution Discharge Elimination System. In 1972, Congress passed sweeping and far-reaching amendments to the water pollution control legislation, known as Public Law 92-500. There were many sections to this legislation, some of which would become operative many years later. Section 201 established a program of assistance to communities for the construction of sewage treatment plants. Section 208 established the concept of areawide management plans for the control of nonpoint pollution sources. Section 402 established the National Pollution Discharge Elimination System (NPDES), which had as its goal the elimination of pollutant discharges. Twenty-one point source categories were established within the NPDES framework, and guidance documents were prepared outlining the conditions under which discharge from each of these source categories would be permitted. Concentrated animal feeding operations were one of the point sources identified and a permit program was established. The program contained provisions that individual states could be certified to administer within the NPDES system but only after they had established the appropriate regulatory structure.

As one of the point sources covered by the NPDES criteria, feedlots were defined and a program established whereby the states were to be granted permitting authority. Within the definitions specified, an animal feeding operation was defined as "a lot or facility where 1) animals have or will be stabled or confined and fed or maintained for a total of 45 days or more in any 12-month period; and 2) crops, vegetation, forage growth, or postharvest residues are not sustained in the normal growing season over any portion of the lot or facility." Two or more animal feeding operations under common ownership are deemed to be a single animal feeding operation if they are adjacent to each other or if they use a common area or system for the disposal of waste. The term *manmade conveyance* means constructed by humans for the purpose of transporting waste. The basic structure of the EPA program for livestock feeding operations is outlined in Table 3.1.

Table 3.2 provides the conversion that is used to calculate the number of animal units in an operation for purposes of determining if a permit is required under the NPDES program. During the years immediately after the issuance of the requirements for livestock feeding operation permits, the EPA was actively involved in permitting efforts and in developing the technologies necessary to prevent water pollution. During the intervening years, various states have been given the authority to grant

Table 3.1. Basic structure of the EPA program for control of pollution from livestock feeding operations

Feedlots with ≥1,000 animal units	Feedlots with <1,000 but with ≥300 animal units	Feedlots with <300 animal units
Permit required for all feedlots with discharges of pollutants.	Permit required if feedlot 1. Discharges pollutants through an artificial conveyance, or 2. Discharges pollutants into water passing through or coming into direct contact with animals in the confined area. Feedlots subject to case-by-case designation requiring an individual permit only after on-site inspection and notice to the owner or operator.	No permit required (unless case-by-case designations as specified below). Case-by-case designation only if feedlot 1. Discharges pollutants through a manmade conveyance, or 2. Discharges pollutants into waters passing through or coming into direct contact with the animals in the confined areas; and after on-site inspection, written notice is transmitted to the owner or operator.

Table 3.2. Number of animals that are equivalent to 300 and 1,000 animal units for purposes of EPA NPDES permit program

Animal	Number equivalent to 300 animal units	Number equivalent to 1,000 animal units
Slaughter or feeder cattle	300	1,000
Mature dairy cattle	200	700
Swine > 55 lb	750	2,500
Sheep	3,000	10,000
Turkeys	16,500	55,000
Chickens with continuously flowing water	30,000	100,000
Chickens with a liquid manure handling system	9,000	30,000
Ducks	1,500	5,000

NPDES permits and the program operates independently of EPA intervention.

State Programs for Control of Water Pollution from Confined Animal Feeding Operations

Individual states have assumed responsibility for the confined animal feeding operation (CAFO) permit programs. In most states, the previously existing water pollution control agency accepted the responsibility.

Table 3.3.   Issues included in typical state CAFO permit applications when a cattle feedlot is being proposed

Parameter	Issue	Assurance sought
Area of the feedlot and other runoff-contributing surfaces	Determines the amount of runoff that can be anticipated from the design storm	That the designer has included all of the runoff collection area so the runoff retention basin will have adequate storage capacity for the location
Volume of the runoff retention basin	Determines the amount of runoff that can be stored before discharge occurs	That the retention basin has sufficient capacity
Surface area of the runoff retention basin	Determines the additional volume that must be stored due to precipitation falling directly on the basin	That provision was included in the design to accommodate this additional liquid volume
Equipment available to remove the collected runoff and apply it to cropland	Determines the capability to remove accumulated runoff in a timely fashion	That adequate equipment is available to keep the volume of stored runoff under control so there is capacity for the next rainfall event
Land area available for the disposal of accumulated runoff	Determines the amount of runoff that will need to be applied per acre	That there is adequate land available
Cropping practice, including whether supplemental irrigation water is to be used, on the land allocated to runoff disposal	Determines if the cropping practice is adequate to use the nutrients applied	That the nutrients, especially nitrogen, will be used by the crop and will not be carried to the groundwater
Nature and extent of the proposed monitoring	Determines if the data collected will be sufficient to alert the operator and the regulatory agency of problems before major damage is done	Is the groundwater being adequately protected
Type and quality of the lining material to be used in the bottom and sides of the runoff retention basins	Determines if there is likely to be downward movement of soluble nutrients	That the basin will be water-tight to the extent required. Typically, a seepage rate of 10^{-7} cm/second is specified

Table 3.3.   *(continued)*

Parameter	Issue	Assurance sought
Distance to the nearest stream, sinkhole, public water supply, well, or other sensitive waters	Determines if there are particular risks inherent in the site that would preclude construction of a feedlot at this site	That the designer has surveyed the site and provided assurance that there is no unanticipated danger
Distance to nearby residences, businesses, and other points of public gathering	Determines if there are likely to be conflicting land uses that will result in conflicts	That the construction of the feedlot is not likely to result in conflicts with existing landowners
Has a public notice of the intention to construct a feedlot been issued	Determines if the approval of this permit application is likely to cause local conflict	That surrounding landowners have had an opportunity to comment on the plan in advance of its approval
What provisions have been included in the design for the control of dust, odors, and flies	Determines if the design process has considered the potential impact on neighbors	That the feedlot planner or designer has included these considerations

In some states, the program generated a new agency or prompted a co-operative unit involving both water quality and agricultural production. The permit application process differs somewhat from state to state; however, in each state there is an overall concern with water quality protection. In some states, there is also an attempt to check for proximity to other development such as housing areas that might suggest conflicts. Table 3.3 summarizes some of the issues that are included in the permitting process. Not all state CAFO permitting agencies currently consider all of these factors nor are all of the factors essential at every location. Table 3.4 is a similar check sheet that can be used in the review of a permit application for a roofed confinement facility. An example might be an application to build a confinement swine unit.

In addition to the permitting process, some states operate a continuing inspection program in which the permitted CAFO is periodically inspected to ensure ongoing compliance. Another alternative is to require monitoring as part of the permit conditions so that the operator provides periodic reports such as water level in the retention structure or the amount of waste applied to particular fields. Some agencies require a nutrient management plan and periodic soil sampling as a way to ensure that the waste is being applied properly. In North Carolina, each animal waste management facility using a liquid system must register and be visited twice per year by a representative of the state regulatory agency. One visit is to check adherence with the Nutrient Management Plan and the

Table 3.4. Issues included in typical state CAFO permit applications when a roofed CAFO is being proposed

Parameter	Issue	Assurance sought
Maximum and average number and type of animals to be housed	Determines the waste load that can be anticipated for the facility	That the designer has included the appropriate calculations to determine the size of the various treatment units
Plan showing the processes to be used in the storage and treatment of the waste	Determines the results that can be expected in terms of organic matter reduction and nutrient loss	That the overall plan is adequate to achieve the promised results
Size and capacity of each of the waste management components	Determines adequacy of the devices and whether they have been designed properly	That provision was included in the design to accommodate the load
Equipment available to remove the stored waste and apply it to cropland	Determines the capability to remove accumulated waste in a timely fashion	That adequate equipment is available to keep the volume of stored waste under control so there is capacity for further accumulation in case rain precludes land application
Land area available for the disposal of accumulated waste	Determines the amount of treated waste that will need to be applied per acre	That there is adequate land available
Agronomic practices, including supplemental irrigation water to be used on the land proposed for application of the waste	Determines if the cropping practice is adequate to use the nutrients applied	That the nutrients, especially nitrogen will be used by the crop and will not be carried to the groundwater
Nature and design of the monitoring program that is to be used	Determines if the data collected will be sufficient to alert the operator and the regulatory agency of problems before major damage is done	That the groundwater is being adequately protected
Type and quality of the lining material to be used in the bottom and sides of any lagoons or earthen storage basins	Determines if there is likely to be downward movement of soluble nutrients	That the basin will be watertight to the extent required. Typically, a seepage rate of 10^{-7} cm/second is specified

Table 3.4. *(continued)*

Parameter	Issue	Assurance sought
Distance to the nearest stream, sinkhole, public water supply, well, or other sensitive waters	Determines if there are particular risks inherent in the site that would preclude construction of a confinement facility at this site	That the designer has surveyed the site and provided assurance that there is no unanticipated danger
Distance to nearby residences, businesses, and other points of public gathering	Determines if there are likely to be conflicting land uses that will result in conflicts	That the construction of the feedlot is not likely to result in conflicts with existing landowners
Has a public notice of the intention to construct a CAFO been issued	Determines if the approval of this permit application is likely to cause local conflict	That surrounding landowners have had an opportunity to comment on the plan in advance of its approval
What provisions have been included in the design for the control of dust, odors, and flies	Determines if the design process has considered the potential impact on neighbors	That the planner or designer has included these considerations

second to check proper operation of the Certified Waste Management Plan. A maximum level indicator has been installed in all lagoons to ensure maintenance of sufficient storage for the runoff from two 25-year, 24-hour storms. Anyone observing a lagoon depth exceeding the maximum allowable level is required to report it to the regulatory agency.

State agencies also may have specific regulations concerning the disposal of dead animals. Among the options for dead animal disposal are trucking to a rendering plant, on-site composting, incineration, or burial. Any one of these options can be done satisfactorily but each requires a degree of management to avoid nuisance and water quality problems.

State Programs for Control of Nonpoint Source Pollution

During the first years of PL 92-500, the emphasis was on the control of point sources of pollution and the permit programs were given major emphasis. More recently, however, state agencies have recognized that nonpoint source pollution is responsible for a significant number of the water quality criteria violations, and they have launched major efforts to reduce nonpoint pollution sources. Regulatory control of nonpoint source pollution is difficult. In many states, the responsibility for these efforts is being shifted to local agencies such as the Soil and Water Conservation districts and watershed councils. Each of these agencies attempts to operate on the basis of education and incentives rather than by the more traditional regulatory approach.

Section 208 of PL 92-500 outlines a scheme of watershed assessments that are to be conducted locally as part of the nonpoint source pollution reduction effort. These watershed plans outline and identify the sources of nonpoint source pollution and then begin to seek remedial efforts to solve issues that can be remedied. For example, Oregon is in the midst of a major effort to reduce nonpoint source pollution as part of a program to restore salmon fisheries; streambank erosion, poor grazing practices, and forest harvest practices have been identified as three of the major pollution contributors. As a result, efforts are underway on a local scale to restore damaged streambanks, to eliminate tree cutting along streams, and to reduce animal traffic in riparian areas. So long as these local, voluntary efforts are effective, they are being supported as more effective and less expensive than traditional regulatory approaches.

Funds are being made available to local groups through a variety of federal, state, and local agencies to promote improved land management with an eye toward nonpoint source pollution control. These locally identified and initiated efforts are considered more effective than attempting to regulate land use practices on a statewide basis.

Mechanisms for Control of Air Quality and Odors

Extensive regulatory attention has been devoted to air quality during the past 30 years. Earliest efforts were devoted to ensuring worker health in their places of employment. These industrial hygiene standards specify the maximum allowable concentration of various materials that may be in workplace air. Normally, these values are established based on an 8-hour exposure and consider the health of the worker. It was recognized quickly that these standards were not applicable to the air quality in a residence located in proximity to a livestock feedlot or other confinement facility.

Worker and Animal Safety Considerations

Most measurements have documented that the air quality within a livestock confinement building or within a cattle feedlot does not violate the workplace air quality standards based on ammonia, methane, or hydrogen sulfide. There is a particular situation, however, in which this is not the case. Early in the history of swine confinement units in the United States and in dairy cattle barns in northern Europe, it was observed that when manure storage tanks were agitated after standing quiescent for a time, animals in the vicinity of the storage tank were dying. And in a few cases, operators were killed by the evolved gases. Most authorities attribute this phenomena to the evolution of accumulated hydrogen sulfide. There was a particularly tragic case in northern Ohio in which two brothers operating a dairy farm were both killed. Investigations seemed to suggest that one brother went to the manure storage tank, started the agitator, and either slipped and fell into the tank or was overcome by gases and fell into the tank. Subsequently, it is surmised that the second brother came to investigate, found his brother in the tank, and attempted to rescue him. As a result, both brothers were killed.

This experience and countless others have led to the understanding that manure storage tanks and aerobic units that have been allowed to stand without aeration more than 2 hours are potentially lethal gas sources when agitated. Several safety precautions follow:

1. All manure storage and treatment tanks should be protected by covers, grates, or handrails that prevent accidental falls into the tank. Children are included in the group to be protected so guards should be designed based on their smaller size.
2. Manure storage tanks and tanks used for the treatment of manure should be designed to eliminate or minimize the necessity of entry as part of normal operation and maintenance. Thus, pumps, agitators, and valves should be serviceable from outside the tank, or they should be removable for service.
3. Tanks in which the operating level is more than 2 feet below the top should be fitted with interior steps or a ladder to allow a person to get out should he or she accidentally fall into the tank.
4. No one should enter a manure storage tank without wearing appropriate safety gear. Self-contained breathing equipment is the most suitable approach. Mechanically ventilating the space above the liquid level prior to entering the tank is of some help and is indicated even when other alternatives are in place
5. When it is time to agitate a tank, animals should be removed from the immediate area, high levels of ventilation provided, and worker safety considered. These precautions are particularly important for tanks located within a confinement building or housed within a protective shed.

Odor Control Considerations

Odors are the most frequent complaint about livestock and poultry confinement facilities. Odors have proven difficult to control from an engineering perspective as well as difficult to regulate. Federal air quality regulations are silent regarding odors. State and local jurisdictions have repeatedly attempted to address the issue, however, because of the difficulty in measuring odors, most of these attempts have proven less than satisfactory. As indicated when discussing water quality protection, the state water quality regulatory agencies have shown some concern over odors in their permitting programs by being aware of homes, businesses, and other public facilities in the proximity of livestock operations. Most of these agencies do not have the authority to firmly deny permits based on odor concerns, but they can raise the issue and assist the developer by bringing these concerns to his or her attention. County and state land use agencies have jurisdiction in some areas and have attempted to regulate the location of livestock facilities based on separation distances from other development. Because livestock enterprises have become larger, separation distances that were adequate in the past have proven insufficient. Part of the challenge is due to the lack of an agreed standard on an acceptable detection frequency of an odor. Similarly, there is not a standard for the intensity or quality that meets the need of both the livestock producer and the neighbor. These complexities are discussed in greater detail in a separate chapter. We can conclude there is no federal, state, or local odor control regulatory process that adequately protects nearby residents while at the same time ensuring the livestock producer that his or her operation will not inappropriately be considered a nuisance.

Without an odor regulatory system, individuals and groups that think their rights have been violated by the odors from a livestock or poultry operation have turned to the court system for relief. Unfortunately, this path is costly and uncertain for both the resident and the accused livestock operation. Most court actions for relief from odors have been taken under the Law of Common Nuisance, a long-standing legal principle that says that an individual or group has the right to enjoy its property without unreasonable interference by the activity of someone else. Thus, residents have the right to enjoy their homes, practice their professions, or pursue their business without unreasonable interference by livestock odors. The difficulty is in establishing what is the limit to reasonable interference. Most people would agree that a single detection of a mild odor once per year is not an unreasonable interference, and they would further agree that the constant presence of a stench that kept friends from coming to visit is unreasonable. In between are numerous shades of reasonableness that are impossible to decide on an objective basis.

Ideally, the neighbors and the livestock confinement system operator would be able to sit together and determine a course of action that would provide an acceptable solution to both parties. There are techniques that reduce odors. Unfortunately, most of the less odorous alternatives cost more or are more difficult to manage than those recognized as major culprits of odor production. Anaerobic manure storage basins are a prime example. Large earthen basins can be constructed that meet the standards for infiltration and hence protect and enhance the groundwater. Unfortunately, they are also major sources of odor. Providing a cover to trap the odors, an aerator to maintain aerobic conditions, or adoption of an alternate system that stores and treats the manure in an enclosed tank is more expensive than the use of an earthen storage basin. This is a dilemma for the designer. The system that is selected will clearly impact the construction and operating cost of the facility, and if the wrong decision is made, may involve even more costly retrofitting at some time in the future. The designer or planner can become better prepared to make these decisions by studying the subsequent chapters of this book; however, it remains a difficult challenge until we as a society better define our standards for odor control in the rural environment.

There is developing evidence that odor can cause stress that increases existing health problems such as respiratory diseases or that it may even be the causative agent for some health problems.

Mechanisms to Control Spread of Diseases from Livestock to Humans

Public health is protected by an intricate system of rules and regulations in the United States that result in a variety of protective measures ranging from chlorination of water to identification of the processing lot in canned foods. Similarly, we are protected from disease due to contact with animal and poultry manure. Most notably, the water quality standards of the states for public waters have a criterion related to the presence of indicator organisms. Indicator organisms are those typically enumerated to detect the presence of human pathogens in water. Fecal coliform, *Escherichia* coliform, and fecal streptococcus organisms have been most commonly used. Although none of these organisms are considered important pathogens, all are excreted in large numbers by humans; hence, their presence in a water supply suggests the possibility of human fecal contamination. Livestock and poultry are also significant sources of these organisms and if allowed to enter a watercourse would quickly increase the counts beyond that allowed by the water quality standards. Such water quality standards have effectively eliminated the discharge of animal wastes in an ongoing manner in the United States.

There are other sources of indicator organisms from livestock production that are not so effectively regulated. For example, runoff from crop and pastureland to which manure has been recently applied is of concern in watersheds of western Oregon where commercial oyster harvesting is a protected water use. In this area, dairy farmers are taking specific measures to reduce organism transport. Grazing animals immediately upstream of a recreational area is another area of concern as is grazing in a public water supply watershed. Each of these situations requires individual consideration.

Manure is typically applied to crop or pastureland as a way to recycle nutrients. When fresh manure is applied, there is the possibility of disease organisms being on the surface of the crop. In a few situations, it is possible for pathogenic organisms on the surface to actually enter the crop, particularly if there is a surface blemish or the stem area is exposed. As a safety precaution, manure tends not to be used on vegetables to be consumed fresh from the field without cooking.

Although infrequent, there is an occasional concern over airborne pathogens from livestock and poultry operations infecting nearby residents. The epidemiological evidence for this concern is slight. Research shows that dusts and aerosols are carried downwind from facilities and from land application sites, however, there is little evidence of this source being a viable mode of disease transmission. It is reasonable, however, not to establish a livestock or poultry confinement facility near a particularly sensitive site where public health risks would result. For example, a poultry operation was proposed close to a tomato processing facility in South Carolina that would seem to pose an unreasonable risk of airborne dusts landing on the tomatoes after their having been washed that could be carried to unsuspecting consumers hundreds of miles away.

Right to Farm Laws and Livestock Production

In the 1980s, farmers in several states became concerned about the number of nuisance suits brought against them based on odors. The concern that prompted this was that normal agricultural practices such as the application of manure from traditional livestock production might become the basis of legal action against a farm family, forcing them into an expensive legal defense. The second factor prompting these laws was the growth of cities and towns into previously rural areas. For example, a cattle feedlot operator in eastern Idaho became the object of a nuisance action after several new houses were constructed within 0.5 mile of the feedlot. The feedlot had been in existence for more than 40 years and thus had developed well in advance of the growing suburbs. In this particular case, there was also a group within the community promoting the construction of a golf course on property not far removed from the feedlot. Had a "Right to Farm Law" been in place, it might have prevented the litigation, however, the problem was more serious. The real issue was,

How do you change a land use when it has become inappropriate over time? It was clearly evident that it was no longer reasonable to have a feedlot in that location, however, the feedlot operator had a sufficiently large investment in the facility that he or she could not easily afford to duplicate at a more remote site for the comfort of the neighbors. This situation was not fully resolved by the introduction of the legal action. The feedlot was allowed to continue on the basis that it was there first, was being well operated, and was consistent with other agricultural and industrial operations in the area.

The Right to Farm laws differ among the states but all attempt to protect established agricultural practices from becoming the basis for litigation. The laws essentially protect established farming practices. From the perspective of livestock producers, the Right to Farm Laws do not provide a license to engage in faulty management that produce elevated odors and flies. The laws, however, do provide a degree of protection to an established agricultural operation if a housing development should be constructed nearby that might pose a threat.

Regulation of Livestock and Poultry Wastes Worldwide

The United States is not the only country in which livestock and poultry production technologies are undergoing rapid change or where it has become necessary to develop management systems to protect the environment. A few examples follow that outline alternate approaches being taken around the world, although at this point, none has been demonstrated to be the universal answer.

Singapore

Singapore, an island nation with a high population density and a strong commitment to environmental preservation, conducted what has become a classic experiment in pig production. Singapore has a large Chinese population and many of the residents have a preference for fresh pork in their diets. Prior to 1978, the pigs in support of this market were grown on small farms throughout the island. The pigs were raised in minimal housing and the manure flushed into receiving streams that flowed to the sea. It became obvious that this practice was not acceptable and that the country needed to protect water quality to support its overall economic development.

It was decided to establish a "pig estate" in the northwestern area of the island, remote from the center of the city and the least densely populated. This area was to be set aside for the development of modern pig farms. Several options were considered for manure management, including areawide treatment with biogas recovery and individual lagoon-based treatment systems followed by algal production or conventional secondary treatment. A major research initiative was launched. Success was achieved in operating an algal basin over an extended time. However, by 1985 the government decided to abandon the pig estate concept and to import pork from other countries in which more land was available to support pig production. The land set aside for pigs in Singapore was used for alternate purposes. One can conclude that it is very expensive to raise pigs in a nonagricultural setting in which there is no land available for manure use and separation from intensive housing development is only a few hundred meters. In the process of seeking an answer to their particular problems, the researchers in Singapore made important contributions to the international understanding of pig waste treatment. It is unfortunate that waste treatment and odor control from the buildings could not be achieved in Singapore.

Taiwan

Taiwan has a large number of pig farms ranging from small family units to a few very large integrated pig farms such as the President's farm, which has more than 100,000 hogs. There is a standard wastewater treatment system for the smaller farms that consists of what they call the "red mud" digester, a rectangular basin covered with a red mud plastic cover in a half-cylinder shape that captures the biogas. Effluent from the digester typically goes into a small aeration basin from which the effluent is discharged into a stream or ditch. Essentially, none of the manure is used as a crop fertilizer. The government has been actively involved in the development and improvement of treatment processes. The current goal is to achieve an effluent biochemical oxygen demand (BOD) of 80 milligrams per liter and a total suspended solids concentration of 200 milligrams per liter. Expectations are that these limits will be reduced to 20 milligrams per liter BOD and 100 milligrams per liter total suspended solids.

Under the leadership of the U.S.–Asian Environmental Partnership, a project is being conducted in southern Taiwan to demonstrate and evaluate alternative swine waste treatment technologies pursuant to solids use and stream discharge. This project serves as the basis for a livestock waste management center. Plans are for this center to provide leadership in educational outreach, training, student and faculty exchange programs, development of advanced waste treatment technologies, and hands-on experience for waste treatment specialists from cooperating countries in Southeast Asia.

Even in a country in which all the petroleum is imported, there is very little interest in capturing plant nutrients because pig farmers generally have no cropland, and adjacent farmers, if there are any, would prefer to purchase commercial fertilizer that has high quality and no odor. There is, however, an interest in the production of compost for use in horticultural nurseries and home gardens.

Korea

In Korea, systems are being developed and implemented for the separation of solids and liquids, with the solids being composted and the liquids being treated for stream discharge. Belts below the slatted portion of the hog pens provide continuous separation of liquids for the liquid treatment system and frequent solids removal as input to composting. The liquids are treated by screening, primary sedimentation, aeration, and final settling prior to discharge. Solids from the screening operation and the settling tanks are added to the composting unit.

Thailand

In Thailand, effluent requirements are being developed for three different sizes of swine operations to recognize that it is more difficult for small-scale farmers to implement new and improved treatment techniques than larger-scale farmers. Effluent limitations for units with fewer than 60 animals are as follows: BOD = 120 milligrams per liter, chemical oxygen demand (COD) = 500 milligram per liter, total Kjeldahl nitrogen (TKN) = 300 milligrams per liter, and TSS = 300 milligrams per liter. For units with 60 to 600 animals, BOD = 100 milligrams per liter, COD = 400 milligrams per liter, TKN = 150 milligrams per liter, and total suspended solids (TSS) = 150 milligrams per liter. For units with more than 600 animals, BOD = 60 milligrams per liter, COD = 250 milligrams per liter, TKN = 100 milligrams per liter, and TSS = 100 milligrams per liter. The governmental energy conservation agency is providing cost share funds for digesters from which biogas is used. Many small land unit operators raise swine on ground or dry lots and have no waste treatment facilities.

Western Europe

Dairy, pig, and poultry units of western Europe more nearly resemble the U.S. counterparts with a strong commitment to return manure to cropland. In addition to the standard U.S. concerns, some of the European countries attempt to regulate the volatilization of ammonia and methane. Regulatory constraints are prevalent and farmers seem to pay attention to manure management. Much of this response can be explained by the smaller farms; greater population density near the farms; and in some areas, higher rainfall and more severe winters than in U.S.

areas. European countries were the leaders in the development of the oxidation ditch, a modified aerobic activated sludge treatment process, and they have used methane digesters to a greater extent than in the United States.

Australia

In Australia, a large swine production company has moved facilities to an interior location where human populations are sparse and thus problems with odor and water pollution would be decreased. However, techniques are being developed to control odor by frequent waste removal, solid–liquid separation, and advanced waste treatment techniques for solid and liquid components.

4

Waste Characterization Relative to Environmental Quality Impacts

Any characterization of livestock and poultry wastes to demonstrate their potential adverse environmental quality impact is fraught with hazards. This is in part because students of this topic generally have a background in municipal or industrial waste management in which wastewater typically pours from a pipe into a receiving stream on a relatively continuous basis. Fortunately, that is not how livestock and poultry wastes are managed in the United States and most of the rest of the world. Livestock do not have the same kind of waste handling or collection systems that are typical for municipal systems. As a result, the practitioner of livestock waste management is prompted to deal with a greater variety of raw materials and to deal with them on a schedule that may be irregular and dictated by conditions such as rainfall, which are beyond control. This book attempts to leave options open for a variety of treatment schemes and innovations. Not all of the workable schemes have been exploited to date and as demands and constraints change, new opportunities and requirements arise.

In this chapter, several typical livestock waste materials are identified. They are, in general, materials that are currently encountered and for which treatment or management is required. Certainly, this list is not all-inclusive nor does it match precisely what is produced at any particular livestock or poultry operation; however, it provides an approximation to many of the materials and identifies many of the options currently in use to prevent environmental damage.

Dry Manure Mixed with Inert Solid Material

Cattle feedlots are designed to hold steers for approximately 6 months while they increase in weight from approximately 600 pounds to a market weight of 1,000 to 1,200 pounds. During this time, they are typically fed twice daily and provided a sufficiently high-energy feed so they gain at their full genetic potential. Manure is allowed to accumulate on the feedlot surface. The action of the animals moving within the pens mixes

43

Table 4.1. Amount and composition of material scraped from a
cattle feedlot or similar earthen feeding area

Parameter	Value
Weight of material	28 lb/day/1,000 lb live weight
Total solids (TS)	56% of wet weight 14 lb/day/1,000 lb live weight
Volatile solids (VS)	4.8 lb/day/1,000 lb live weight
VS/TS	38%
BOD_5	5.3% of wet weight 0.75 lb/day/1,000 lb live weight
Total nitrogen	1% of wet weight 0.24 lb/day/1,000 lb live weight
Ammonia nitrogen	0.35% of TS 0.038 lb/day/1,000 lb live weight
Total phosphorus	0.65% of TS 0.071 lb/day/1,000 lb live weight
Potassium	2.0% of TS 0.22 lb/day/1,000 lb live weight

Source: Adapted from Overcash et al. 1983.

fresh manure with the previously deposited material and any soil that is
dislodged from the pen base. The pens are groomed on an as-needed
basis with a tractor blade similar to that used to maintain unpaved roads.
In many areas of the country, accumulated manure solids are pushed
into a pile so the animals have a mound on which to rest during times
when the lot is wet or muddy. Animals also tend to stand on the mounds
during hot weather, seeking a cooler spot.

Periodically, it is necessary to enter the lot with appropriate equipment, typically a front-end loader, and remove the accumulated manure
solids, restoring the feedlot surface to its original slope. Manure removal
may be practiced between each batch of cattle or may be scheduled once
a year. Where lots are especially flat or where there are low spots, which
accumulate excess water, more frequent manure removal may be necessary to maintain an acceptable feedlot surface for optimal animal performance. Table 4.1 provides typical values for the material scraped from
feedlot surfaces. There is certain to be wide variability in this material
and if precise application rates are needed, analysis of the actual material becomes necessary.

Table 4.1 indicates that the material is indeed manure and has sufficient organic material and nutrients to have value when land-applied.
But compared with fresh beef cattle manure (Table 4.2), it has picked up
considerable inert material from the feedlot and has lost a considerable
amount of its original nitrogen concentration. The ammonia is almost all

Table 4.2.   Amount and composition of fresh beef manure

Parameter	Value
Weight of material	52 lb/day/1,000 lb live weight
Total solids (TS)	15% of wet weight 7.1 lb/day/1,000 lb live weight
Volatile solids (VS)	6 lb/day/1,000 lb live weight
VS/TS	81%
BOD_5	2.5% of wet weight 1.4 lb/day/1,000 lb live weight
Total nitrogen (N)	0.59% of wet weight 0.39 lb/day/1,000 lb live weight
Ammonia N	4.5% of TS 0.32 lb/day/1,000 lb live weight
Total phosphorus	0.97% of TS 0.07 lb/day/1,000 lb live weight
Potassium	2.0% of TS 0.2 lb/day/1,000 lb live weight

Source: Adapted from Overcash et al. 1983.

gone, having volatilized while the manure was on the feedlot surface. Most cattle feedlot operators haul this material to cropland immediately upon removal from the lot or it may be stockpiled for later distribution when it can be more effectively incorporated into the cropping scheme. Another alternative is to use the material in a compost-making operation. There is at least one feedlot located near a paper processing plant in which the feedlot manure is combined with paper waste to produce compost that is in local demand.

In addition to having lost a significant fraction of the organic matter and nitrogen, the feedlot manure has undergone a considerable decrease in indicator organisms due to its lengthy exposure to sunlight and aerobic conditions. The data suggest that fresh beef cattle feces has an *Escherichia coli* concentration of approximately 3.6×10^9 per gram of dry solids, whereas the concentration in the manure removed is approximately 1×10^6 viable cells per gram of dry solids. This represents a reduction of more than 99.9%.

The data presented in Table 4.1 are specifically based on feedlot scraping samples, however, they can be used in estimating the characteristics of scrapings from other manure-covered surfaces should that information be needed. These characteristics are not appropriate for estimating the quality of poultry litter or other types of manure mixed with bedding.

Management concerns with feedlot manure are to prevent its being washed into receiving streams either during storage or at the time of application to cropland. In addition to its significant biochemical oxygen demand content, it is a source of viable indicator bacteria. Thus, if it is stored on site, the storage area needs to be included in the area from which runoff is collected rather than bypassing the runoff retention structure. As a source of plant nutrients, the feedlot manure is deficient in nitrogen compared with phosphorus and potassium. This means that if applied to meet the nitrogen needs of the crop, excess phosphorus and potassium are applied. If it is applied according to the phosphorus recommendations, supplemental nitrogen is likely to be required.

Runoff from Cattle Feedlots and Other Manure-Covered Surfaces

Runoff from manure-covered surfaces is an intermittent source of livestock waste and as such requires a different kind of analysis than more nearly continuous sources. Feedlots are most frequently built in regions of low rainfall so a runoff collection system is constructed in anticipation that it may be used only a few times per year and not at all during some years. Areas differ in their rainfall patterns. Some regions are characterized as having infrequent but high-intensity thunderstorms that produce large amounts of runoff in a short period. Other areas have more frequent but smaller rainfall events. Thus, in planning a runoff control system for a feedlot or other manure-covered area, it is important to carefully consider local weather data.

Quantity

Runoff from cattle feedlots was one of the first forms of livestock waste to be brought to public attention as a potent source of water pollution capable of causing fish kills and other severe water quality problems. The process is relatively straightforward. As rain falls on a manure-covered surface, the initial portion is adsorbed and trapped in the manure. Thus, there is a delay between the time when rainfall begins and when the first runoff occurs. If rainfall stops during this period, there is no runoff. Assuming rain continues beyond the surface holding capacity of the lot, runoff begins first from the points closest to the outlet then progressing up the lot until finally runoff represents a contribution from the total surface. The amount of runoff from a particular storm can be predicted with the classical Soil Conservation Service runoff equation (Schwab et al. 1966).

$$Q = \frac{(P - 0.2S)^2}{P + 0.8S}$$

$$S = \frac{1,000}{N} - 10$$

where Q = surface runoff (in inches); P = rainfall (in inches); S = maximum potential difference between rainfall and runoff (in inches); and N = an empirical number characterizing the runoff producing surface. A value of $N = 100$ would describe a surface devoid of storage. A value of $N = 91$ has been found to describe a dry feedlot surface, and a value of 97 a wet lot.

The amount of runoff that can be anticipated from feedlots at various locations is related to the average annual precipitation and to the distribution of the precipitation. Figure 4.1 shows the 24-hour rainfall that can be expected to occur at a frequency of once every 25 years at various U.S. locations. For example, Pendleton, Oregon, receives slightly more than 13 inches of rainfall per year, but most of it falls in small storms so only approximately 12% of the rainfall runs off the lot. In contrast, Ames, Iowa, is more likely to receive larger storms, hence 36% of the rainfall must be managed as runoff. This information is summarized in Table 4.3. From this table, it is clear why feedlots tend to be built in areas where there is less runoff to be captured and managed.

Table 4.3. Average annual runoff from earthen feedlots at seven representative locations

Location	Average annual rainfall, in.	25 Year - 24 hour rainfall, in.	Average annual feedlot runoff, in.	Runoff to rainfall ratio
Pendleton, OR	13.4	1.5	1.60	0.12
Lubbock, TX	18.6	5.0	5.99	0.32
Boseman, MT	19.2	2.7	4.76	0.25
Ames, IA	30.9	5.4	11.05	0.36
Corvallis, OR	39.7	4.5	12.52	0.32
Experiment, GA	49.9	6.7	19.40	0.39
Astoria, OR	75.4	5.5	32.7	0.43

Quality

Shortly after it was discovered that runoff from feedlots was responsible for severe water pollution incidents, efforts were initiated to determine the characteristics of the runoff. Runoff characterization studies were conducted in several Great Plains states. Immediately, it was noted that the quality was highly variable but in all cases was sufficiently concentrated in both manure particulates and soluble extracts to require retention. Feedlot runoff characteristics are summarized in Table 4.4, which includes both representative values and ranges.

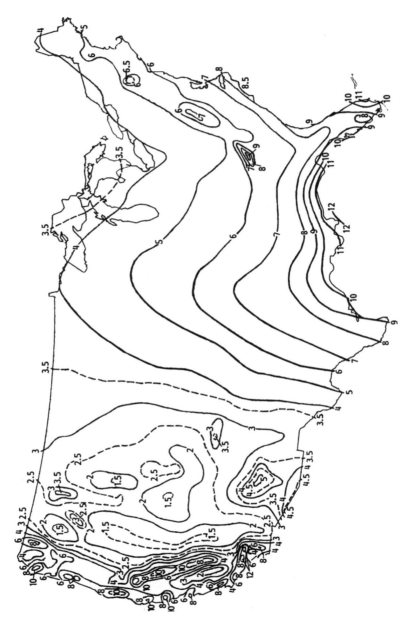

Figure 4.1. The 25-year, 24-hour rainfall amount at various locations around the United States.

Table 4.4.   Representative values for the quality of rainfall initiated cattle feedlot runoff and the runoff from other manure-covered surfaces

Characteristic	Typical value	Normal range
Chemical oxygen demand, mg/l	10,000	3,000–30,000
Biological oxygen demand, mg/l	2,500	800–10,000
Total solids, mg/l	15,000	5,000–40,000
Volatile solids, mg/l	5,000	300–20,000
Total nitrogen, mg/l	350	100–2,000
Ammonia nitrogen, mg/l	150	10–1,000
Phosphorus, mg/l	300	10–500

Runoff from feedlots is also likely to carry a relatively large concentration of indicator bacteria. Indicator bacteria are bacterial groups that are typically enumerated to indicate the sanitary quality of water. They are generally of less concern from a public health perspective than specific disease-producing organisms but are counted because the procedures are easier, quicker, and less expensive. The presence of indicator bacteria is taken as an indication that pathogenic species are likely to be present. Because of the organic strength involved and the usual management techniques used, there have been limited attempts to quantify indicator organisms. The data that have been collected show that total coliform, fecal coliform (FC), fecal streptococcus (FS), and *Escherichia* coliform (*E. coli*) concentrations are likely to be in excess of a million viable organisms per milliliter. Concentrations 10 to 100 times this number have been recorded.

Factors Influencing the Quality of Feedlot Runoff

With the wide variability in the concentration of pollutants in cattle feedlot runoff, there has been an ongoing interest in identifying the factors that are within the control of the feedlot designer and operator to influence the quality. These factors are identified in Table 4.5 in a qualitative manner to guide the design process. The data are so variable it is not possible to provide quantitative predictions.

Environmental Quality Implications

Runoff from manure-covered surfaces is a mixture of soluble and particulate material. Based on the concentration of organic material in runoff, it is capable of rendering large quantities of clean water unsuitable for many uses and fish life. Thus, feedlots of 1,000 head or more that are subject to the National Pollutant Discharge Elimination System permit process must have the ability to capture the runoff, retain it in some type

Table 4.5. Factors influencing the concentration of pollutants in runoff from cattle feedlots and other manure-covered surfaces

Factor	Impact on runoff quality
Feedlot surface material	The processes whereby rainfall striking a feedlot surface mixes with the manure, extracting a portion of the soluble material and eroding a portion of the particulates, is complex and not fully described. It has been noted, however, that runoff from concrete and other solid surfaces is more concentrated than runoff from dirt lots where animal traffic creates a rough surface with a large number of small settling opportunities.
Feedlot slope	Feedlot slope is related to the amount of time water spends on the feedlot surface, hence the amount of extraction of soluble material that is carried with the runoff. Slope also influences the velocity of liquid moving across the surface and hence the ability to transport particulate materials. These two phenomena are clearly in opposition. As a result, the data available do not support an objective relationship between feedlot runoff quality and slope.
Rainfall intensity	The runoff process includes a major component of extraction between water as rainfall and an abundant supply of manure. Data suggest that more concentrated runoff is produced during low-intensity rainfall. Under this condition there is more time for extraction of soluble material but there is a lesser tendency to transport particles by overland flow.
Rainfall duration	The first runoff from a feedlot surface tends to be of lower concentration because it is water that fell near the discharge point and hence had less time in contact with the manure. Runoff concentrations then increase to a maximum as water that has traversed the full lot approaches the outlet. Subsequently, the concentration decreases as "flow ways" or "rivulets" develop in the lot surface, allowing a portion of the runoff to pass across the lot with minimal mixing with the manure.
Lot moisture content	Feedlot runoff characterization data show that runoff from a previously dry lot is less concentrated than that from a lot that was wet prior to the onset of precipitation. This suggests that the extraction of soluble materials from manure is time-related. By maintaining lots as dry as possible and preventing the overflow of watering devices, runoff concentrations can de minimized.

Table 4.5. *(continued)*

Factor	Impact on runoff quality
Frequency of lot cleaning	At various times, management suggestions have been made to reduce feedlot runoff concentrations by adopting more frequent cleaning schedules. In general, the data indicate the lot must be cleaned at least weekly if this practice is to make a perceptible difference. As a result, most designers do not include frequent cleaning as part of the runoff management scheme.
Temperature	Like most processes based on extraction and solution, runoff quality is temperature-dependent. Runoff events during warm periods tend to produce more concentrated runoff than those when it is cooler.
Source of water	There is a special case involved in feedlot runoff that is worthy of mention. If a feedlot has accumulated a heavy load of snow and ice, this material becomes mixed with the feedlot manure. When this mixture of manure, ice, and snow melts there frequently follows an overland flow some have likened to a lava flow after a volcanic eruption. This wave of semisolid material moving down a feedlot may be considerably more concentrated in manure solids than the range suggested in Table 4.4.

of storage facility, and finally to dispose of or use it in an environmentally sound manner. Most feedlots base their systems on applying the collected runoff to cropland by using irrigation equipment.

Feedlot runoff not captured enters a receiving stream where it pollutes the stream and moves with the current, causing fish to die until the oxygen demand is finally satisfied and the stream recovers. For a large feedlot, this could be in excess of 50 miles.

Feedlots smaller than 1,000 head or those otherwise not subject to the NPDES permit requirements may adopt a runoff capture system similar to the larger lots or may select an alternate system. For example, a small feedlot might construct a solids setting pond to capture the larger settleable solids then apply the liquid portion by a gravity distribution system to a pasture or other vegetated area to allow the liquid to be absorbed and the nutrients used. A second option is the construction of a grassed waterway system in which the liquid is filtered and treated by flowing through a densely vegetated area. Both of these options have been approved for use by smaller feedlots where they could be constructed less expensively than conventional retention basins and irrigation systems.

Application to Other Livestock and Poultry Species

Although beef cattle are the principal species raised in the United States on unroofed, open feeding areas, they are not the only ones. Dairy cattle in the southwestern United States typically spend a large fraction of their days in open feeding pens. These pens may be fitted with a shade structure to provide protection from heat but otherwise the system resembles a feedlot. Pigs also are raised under similar conditions. For all these animals, runoff is a liquid sufficiently laden with manure constituents to require capture and management.

There is one other application of this approach. When manure as a solid is removed from either outside pens or from roofed confinement, it may be immediately hauled to cropland or it may be more desirable from either an agronomic or an environmental perspective to stockpile it for application at a later date. The area in which manure is stockpiled, if it is not roofed, should be treated as a contaminated runoff-producing area and the runoff treated similarly to that from a feedlot.

Manure Scraped from Roofed Confinement Facilities or Deposited into Underfloor Storages

There are numerous systems in which animals are housed in a roofed enclosure and the manure scraped from the housing area by a slow-moving scraper system on the floor of the housing. Alternatively, a portion of the housing area may have a slotted floor through which manure falls into a shallow gutter. The gutter may be fitted with a mechanical scraper that is operated one to three times a day, moving the manure into a storage facility. The storage facility may be an above- or underground tank or an earthen basin. There are also scraper systems in which the manure is pushed into a cross-collection gutter where it is hydraulically transported to a liquid facility.

Dairy cows, beef cattle, and pigs are all produced in systems that use manure scrapers. The advantages of these systems are that they minimize the volume of waste material to be stored or processed, and they generally require less labor than systems in which bedding is added to the housing area to absorb the liquids from manure. A disadvantage is that the scrapers generally leave a film of manure behind after passing, hence they tend to be more odorous than some other building designs. Also, the scraping mechanisms are mechanical in contact with the manure so maintenance can be demanding, and the semisolid manure collected may require either drying or the addition of water to facilitate final disposal. Most scraped manure is applied to cropland.

An alternative to the scraping system that generates similar manure is the underfloor storage system. Numerous underfloor storage buildings were constructed when confinement livestock production was initially becoming popular. The idea in such a facility is to have sufficient underfloor storage capacity to hold the manure for application twice a year.

Characteristics of Manure Scraped from Buildings or Stored in Underfloor Pits

Tables 4.6–4.8 summarize typical characterization data for manure scraped from dairy, beef, and swine confinement units. Note that these are typical data and may not accurately reflect the manure from a particular operation. These tables were prepared assuming a minimal amount of water being allowed to enter the scraping system and a minimum of feed being wasted. If this is not true, the values should be adjusted accordingly. What is most striking about the data in these three tables is the similarity of the three types of manure, thus they probably can be handled in similar systems.

Manure that is scraped from buildings or allowed to accumulate in pits beneath slotted floors has the least volume of the various forms, hence can be stored in the smallest possible storage volume. In addition, if it is to be hauled to cropland or pastures in either a tank wagon or a manure spreader, this minimal volume is a distinct advantage. The disadvantage of this form of manure is that it is too thick to flow well and too thick to be pumped with conventional centrifugal pumping equipment. Diaphragm and piston style pumps are available to handle this material, however, they are generally more expensive than centrifugal pumps.

Environmental Quality Implications

Manure scraped from a building without significant amounts of bedding is in the most concentrated form, and if destined for cropland application has maximum nutrient content. Clearly, it is a concentrated organic waste and should not be allowed to escape into the environment. Similarly, because it is in a concentrated form, most livestock producers elect to protect it from water addition. Rainfall additions only dilute the material. Runoff from a storage area for manure of this type would be heavily contaminated and should be treated the same as feedlot runoff.

In addition to being a concentrated semisolid organic waste, the manure scraped from an operation is an active biological material that evolves odorous gases almost immediately. As a result, storage in pits beneath slotted floors raises the issue of ventilation system design to prevent the accumulation of these gases in the animal environment. Although removal of stored manure from underfloor pits is discussed in detail in a subsequent chapter, it is appropriate herein to mention that

Table 4.6.   Characteristics of manure scraped from dairy housing units. Quantities are expressed as units per day per 1,000 lb live weight

Parameter	Typical value
Volume, gal	12
Total mass, lb	100
Total solids (TS), % wet weight	8
Volatile solids, % TS	85
Total nitrogen (N), % TS	3.0
Ammonia N, % TS	1.8
BOD_5 , % TS	12
Phosphorus, total Phosphorus, % TS	0.76

Table 4.7.   Characteristics of manure scraped from beef cattle housing units. Quantities are expressed as units per day per 1,000 lb live weight

Parameter	Typical value
Volume, gal	7
Total mass, lb	65
Total solids (TS), % wet weight	15
Volatile solids, % TS	85
Total nitrogen (N), % TS	3.0
Ammonia N, % TS	1.6
BOD_5 , % TS	25
Phosphorus, total Phosphorus, % TS	0.9

Table 4.8.   Characteristics of manure scraped from swine housing units. Quantities are expressed as units per day per 1,000 lb live weight

Parameter	Typical value
Volume, gal	8
Total mass, lb	65
Total solids (TS), % wet weight	8
Volatile solids, % TS	80
Total nitrogen (N), % TS	7
Ammonia N, % TS	3.5
BOD_5 , % TS	30
Phosphorus, total Phosphorus, % TS	1.5

when stored manure is disturbed as in mixing to facilitate removal, large quantities of reduced gases, including ammonia and hydrogen sulfide, are released. This release poses an obvious safety concern both for the animals and workers who may be in the area.

Flushed Manure Systems and Other Dilute Manure

In an effort to reduce manure-handling labor and improve building environments, manure flushing systems were developed to serve most species of animals in confinement. Dairy barns are flushed in many parts of the United States as are swine facilities. There have been a few beef confinement systems constructed with roofed barns and flushed gutters. Flush systems are slightly less popular in the poultry industry but they exist. The typical swine or beef flushed system consists of pens arranged over a gutter approximately 18 inches in depth with a slope of approximately 1%. One to four times per day, a flush of water is introduced at the upper end of the gutter and flows to the lower end, carrying the manure that accumulated since the previous flush. This system was among the first attempts to provide a high degree of odor control within the building. Some swine finishing buildings have the gutter in the pen and built so the animals have access to the gutter. This type of gutter is shallow, 2 to 4 inches in depth.

Flushed poultry systems are built with the birds held in cages that are suspended over the gutters. Droppings fall into the gutters and are flushed out as needed. The advantages of these flush systems are low manure-handling labor because all of the hydraulics can be automated, and the wastewater produced is subject to treatment with conventional wastewater treatment equipment. Frequently, the flushing is done with treated wastewater to reduce the amount of fresh water required and to reduce the amount of effluent requiring final disposal. In some parts of the world, this wastewater is treated for discharge into streams. This practice has not been popular in the United States where cropland has generally been available for liquid manure use and disposal; however, there is new and growing interest in systems with effluent discharge because land for manure application is becoming less available and public criticism and regulations continue to intensify to phase out lagoon and spray irrigation systems. Legislation in North Carolina currently requires development of a plan to phase out animal and poultry lagoons.

Flushed manure whether from dairy cows, beef cattle, swine, or poultry is generally between 1 and 2% total solids and except for the additional water, generally reflects the characteristics of fresh manure. Tables 4.9–4.12 provide typical characteristics for flushed manure from dairy, beef, swine, and poultry. Another dilute animal waste source is that

Table 4.9.   Characteristics of flushed manure from dairy barns. Based on using 90 gal of flushwater per 1,000 lb live weight, approximately 100 gal per cow per day. Estimated concentrations for other flushwater volumes can be calculated

Parameter	Value
Volume of wastewater, gal/day/1,000 lb live weight	100
Total solids (TS), lb/day/1,000 lb live weight	10.7
TS, % wet weight	1.2
Volatile solids, % TS	86
BOD_5, % TS	12
Total nitrogen (N), % TS	4.0
Ammonia N, % TS	1.8
Phosphorus, % TS	0.8

Table 4.10.   Characteristics of flushed manure from roofed beef cattle confinement facilities. Based on using 90 gal of flushwater per 1,000 lb live weight, approximately 70 gal per steer per day. Estimated concentrations for other flushwater volumes can be calculated

Parameter	Value
Volume of wastewater, gal/day/1,000 lb live weight	100
Total solids (TS), lb/day/1,000 lb live weight	7.1
TS, % wet weight	0.9
Volatile solids, % TS	81
BOD_5, % TS	26
Total nitrogen (N), % TS	4.0
Ammonia N, % TS	1.5
Phosphorus, % TS	0.9

typically flushed from the dairy milking parlor and associated holding area. Animals are generally moved into the holding area immediately prior to milking and are maneuvered into the milking parlor as the previous group exits. Large quantities of water are generally used to flush the milk lines and clean the milking equipment at the end of the milking session. Frequently, that clean-up water is captured and used to flush the holding area after all the animals have left. Thus, this waste is a mixture of milk, cleaning solutions, and manure from the parlor and holding area floor. The holding area also may be equipped with automatic teat-washing equipment, which makes a slight contribution to the overall waste volume. Table 4.13 contains typical characteristics for this waste source.

Table 4.11.   Characteristics of flushed manure from swine confinement facilities. Based on using 90 gal of flushwater per 1,000 lb live weight, approximately 15 gal per pig per day. Estimated concentrations for other flushwater volumes can be calculated

Parameter	Value
Volume of wastewater, gal/day/1,000 lb live weight	100
Total solids (TS), lb/day/1,000 lb. live weight	6.0
TS, % wet weight	0.72
Volatile solids, % TS	80
BOD_5 , % TS	40
Total nitrogen (N), % TS	6.5
Ammonia N,% TS	3.2
Phosphorus, % TS	1.2

Table 4.12.   Characteristics of flushed droppings from poultry (layers) barns. Based on using approximately 90 gal of flushwater per 1,000 lb live weight, approximately 0.36 gal per bird per day. Estimated concentrations for other flushwater volumes can be calculated

Parameter	Value
Volume of wastewater, gal/day/1,000 lb live weight	100
Total solids (TS), lb/day/1,000 lb live weight	17.8
TS, % wet weight	2.1
Volatile solids, % TS	74
BOD_5 , % TS	20
Total nitrogen (N), % TS	6.1
Ammonia N, % TS	2.6
Phosphorus, % TS	5.3

Tables 4.9–4.12 show that flushing water volumes of approximately 90 gallons per thousand pounds live weight does not give uniformly concentrated wastewaters but in every case provides a liquid manure that can be handled by using liquid manure handling facilities. Currently, there are two popular approaches to handling flushed manure. One is to immediately discharge the manure into an anaerobic lagoon for treatment and storage. The second is to pass the flushed material over or through some type of solid–liquid separator, harvesting the solids, and treating the remaining liquid in one of several ways, depending upon the ultimate disposal and the environmental constraints of the location.

Table 4.13. Typical characteristics of flushed milking
parlor and cow holding area wastes expressed as 1,000 lb
live weight per day

Parameter	Value
Volume of wastewater, gal/day/1,000 lb live weight	25
Total solids (TS), lb/day/1,000 lb live weight	1.2
TS, % wet weight	0.6
Volatile solids, % TS	80
BOD$_5$, % TS	45
Total nitrogen (N), % TS	5
Ammonia N, % TS	2.6
Phosphorus, % TS	2.3

One of the characteristics of the liquid flush system for removing manure from confinement facilities is that it generates a large volume of
wastewater. If climatic conditions and cropping practices allow irrigation
throughout the year, and there is adequate land available, many operators find this is not a problem. In other locations, there is a concern with
this large volume of wastewater because of treatment costs, water availability, and storage volume required if winter application is environmentally unacceptable or if cropping practices preclude application during much of the summer. Thus, systems are developing that retain the
frequent removal and low-labor benefits of the flush system, but that require less water, hence generating less wastewater for treatment. One of
these systems is the pit recharge system in which approximately 6 inches
of freshwater or treated wastewater is added to the level bottomed pit immediately after emptying, allowing manure to accumulate for 3 to 7 days.
The pit is then drained and the material treated as a flushed manure.
This system generates wastewater with up to twice the concentration of
pollutants as conventional flushing but with only half the volume of water to be treated. If eventual discharge to a stream is probable, having a
lesser volume to treat would significantly reduce the cost of treatment.
Pit precharge systems can be automated similarly to flush systems. The
advantages of this option are that the waste volume is less than that of a
flushing system and the odor control benefits of flushing are conserved.

Environmental Implications

Dilute manure slurries such as those from flushed systems offer wastewater
treatment and management options not available with other systems, including separation of liquid and solid fractions either to reclaim a by-product from one or both of the components or to provide a starting material

that can be treated for eventual stream discharge. Flushed systems are popular with many livestock and poultry producers because they generally have lower labor requirements due to the possibility of automating the pumping systems and flushing devices. The large volume, however, means that if this material is to be retained for eventual application to cropland without other treatment, large storage volumes are required. These larger liquid manure storages are generally built as lagoons in the United States.

Lagoons are large earthen basins. Ammonia volatilization, odor, and occasional overflow problems have made them controversial in many areas. They tend to generate odors that many neighbors consider undesirable and if the lagoon is large (some are in excess of 20 acres), the odor is transported on a regular basis to a large area. Another issue with lagoons is related to their risk of failure. Even though a lagoon may be large and constructed beside a stream, the problems should normally be minimal. If, however, during a large storm or other unanticipated climatic event, the lagoon water level overtops the dike and erodes an opening that causes the lagoon to drain, major environmental damage results. Unfortunately, there have been examples of lagoon failure in the United States. Some were related to unusual climatic events, but others were related to operational failures in which the lagoon water level was allowed to rise too close to the overflow or the dikes were not properly maintained. These failures resulted in a sufficiently large number of serious pollution events to cause large lagoons to become unacceptable in some areas. The implied challenge to designers and operators is to do a more reliable job of facility design and construction.

Lagoons can be designed and operated so they do not overflow nor overtop their dikes. Water quality regulations require that lagoons be designed to contain all of the wastewater discharged into them plus have sufficient available capacity to retain all rainfall-induced runoff up to the 25-year, 24-hour storm. Although discharge is allowed for larger storms, virtually no guidance is available concerning how to handle the inflow from these "catastrophic" rainfall events or chronic rainfall over a short time, which exceeds the 25-year, 24-hour storm. In North Carolina where lagoon overflows and embankment breaches have occurred, the state National Resource Conservation Service standards specify that an emergency overflow facility is to be installed to safely allow discharges for rainfall events greater than the legally mandated control criteria.

Another pollution implication of dilute manure systems is the ease with which escape of manure can occur. If solid manure is handled, failure to properly manage the manure generally results in an inconvenience or an expense to the livestock producer but it has limited if any impact on the environment. In contrast, failure of a liquid manure system frequently results in the escape of manure in a form that quickly moves to the closest waterway. This raises a classical design issue. To what extent does the designer include provisions to protect the environment from operational errors and inattention? Most investors are reluctant to pay for facilities that are backup or redundant to cover for operator errors. North Carolina regulations make it mandatory to report any lagoon that has exceeded its allowable liquid level as shown on a liquid level indicator and therefore does not have sufficient storage capacity available to retain the runoff from a subsequent 25-year, 24-hour storm.

A third issue raised by the dilute manure systems is the benefit versus the risks associated with very large livestock operations. The trend of livestock production in the United States and in several other countries has been to have large integrated operations. Dilute liquid manure handling systems are consistent with their size and bring economies of scale in the provision of irrigation equipment and other transport devices. The pattern has been for the design of large anaerobic lagoons to serve these enterprises. These large anaerobic lagoons are significant sources of odor. Also, if a lagoon dike fails either due to extreme rainfall or operator error, a very large quantity of liquid manure is likely to escape with the potential for devastating downstream impact. Thus, by size alone the implications are changed from a local damage to a regional one with many more impacts over time and distance.

Solid Manure Handling Systems

In the past, most manure handling systems used bedding to absorb moisture and produce a solid manure consisting of fecal matter, urine, and bedding material (e.g., straw, wood shavings, or any other locally available, low-cost material) that could absorb moisture and provide a comfortable environment for the animals. The classic system had the advantage of producing manure that could be hauled at the convenience of the farmer, was a useful soil amendment, and had an odor acceptable to the rural countryside. There are currently a number of dairy farmers that use bedding as part of their animal housing. Rather than manually loading the manure from stalls, most of them currently use a tractor with a scraper blade or a front-end loader to handle the manure, or they may use an automated scraper system to move the soiled bedding from the barn to a storage area. The storage area may be covered if it is located in a high-rainfall area. Bedded manure is in demand in some areas for incorporation into compost for eventual use in lawn and garden development. Composted manure also has application in land reclamation projects.

There are other systems producing solid manure. Poultry, especially chicken broilers and turkeys, are commonly raised on litter. A typical process is to place 6 to 12 inches of litter on the floor prior to the first flock of birds and allow the birds to grow to market size. Between flocks, the litter is stirred with a rotating tiller to mix and aerate it prior to the introduction of the next flock of birds. This stirring promotes a composting process that heats sufficiently to control disease transmission between flocks, drives out sufficient moisture to restore the absorptive capacity of the litter, and provides a satisfactory alternative to fresh litter. Sometimes, a small amount of fresh bedding is added. The litter is typically removed annually, hauled to cropland, and replaced with new material. Litter management is critically important in these systems. If the litter is allowed to become too dry, it becomes a dust source, which adversely impacts the birds and workers in the building. If the litter becomes too wet, anaerobic conditions develop and ammonia is released, which is detrimental to the birds, workers, and the surrounding environment due to odor release.

Layer operations for egg production often use an alternative dry manure system called a high-rise house. In this system, a tall building is constructed with an operating floor placed 12 or more feet above ground level. The layers are frequently in cages arranged in a stair-step configuration on the upper level. Droppings fall from the cages to the storage area below. When operating successfully, there is enough air movement through the lower area to dry the droppings to a moisture content below that which supports anaerobic decomposition. Desirable moisture content is in the range of 40 to 50% to avoid foul odors and to provide an accumulated manure that can be removed once or twice annually. Buildings are typically constructed with concrete floors on the lower level and one or more access doors large enough to clean the pit area with a front-end loader. When operating well, these buildings produce high-quality solid manure that can be used as a soil amendment or incorporated into compost as a quality nitrogen source.

A belt system has been used in broiler and layer houses to remove droppings from elevated poultry cages. Techniques have been developed to use this belt system to also move broilers for collection at the end of the belt. Advantages are frequent waste removal for odor and moisture reduction and easier broiler and possibly turkey harvesting.

Tables 4.14–4.16 provide typical solid manure characteristics for different livestock and poultry production systems. Like other manure characterization data, these data are representative and should not be used where precise composition data are required.

Table 4.14.   Composition of dairy cattle manure from bedded housing

Parameter	Value
Total mass, lb/day/ 1,000 lb live weight	81
Total solids (TS), lb/day/ 1,000 lb live weight	15
Total nitrogen, % TS	2.5
Total phosphorus, % TS	0.6
Potassium, % TS	2.3

Source: Adapted from Overcash et al. 1983.

Table 4.15.   Composition of litter and manure removed from poultry houses

Parameter	Value
Mass, lb/day/ 1,000 lb live weight	28
Total solids (TS), lb/day/ 1,000 lb live weight	20
Total nitrogen (N), % TS	3.5
Ammonia N, % TS	1.0
Total phosphorus, % TS	0.6

Table 4.16.   Composition of used litter removed from poultry houses in which the birds were maintained on a litter-covered surface

Parameter	Value
Mass, lb/day/ 1,000 lb live weight	40
Total solids (TS), lb/day/ 1,000 lb live weight	32
Total nitrogen (N), % TS	3.5
Ammonia N, % TS	0.8
Total phosporous, % TS	0.6

References

Overcash, Michael R., Frank J. Humenik, and J. Ronald Miner. 1983. Livestock Waste Management. Boca Raton, FL: CRC Press.

Schwab, G.O., R.K. Frevert, J.W. Edminister, and K.K. Barnes. 1966. Soil and Water Conservation, 2nd Ed. New York: John Wiley & Sons.

5

Livestock and Poultry Production Schemes

Livestock and poultry produce manure whether they are raised on the open range or in confinement. The pollution potential is highly dependent upon the system under which the animals are raised. There are minimal-impact systems in which only a small fraction of the manure reaches a watercourse and other sufficiently dispersed systems and this amount is within the assimilative capacity of the watercourse. The other extreme would involve manure from a confinement system in which all the manure is pushed or flushed into a nearby watercourse. Unfortunately, there have been examples in which this full impact has been noted. Perhaps even more extreme are those unfortunate examples of the recent past in which manure storages have failed, discharging several weeks of accumulated manure or lagoon contents into a stream over a short period. Animal waste spills are more frequent in the United States than necessary. A 1997 Senate study reported there were more than 40 animal waste spills killing more than 670,000 fish in Iowa, Minnesota, and Missouri.

Livestock and poultry are produced throughout the world under an astounding variety of conditions. Cattle, sheep, and goats are raised in many parts of the world by shepherds moving their flocks to useable forage and water. Under these conditions, manure may be allowed to remain where dropped or may be allowed to dry in place and be collected for use as a heating or cooking fuel. Chickens are allowed to forage in lawns and gardens, essentially living off the land in parts of Africa. Although this practice is less common in the United States, it is practiced and was common prior to the development of confinement poultry production in the 1940s. These systems are not particularly efficient nor do they produce the quality of meat that has become the norm in the developed world, but the animals are produced with minimal environmental impact.

Range Production

More than half the beef cattle in the United States are on rangeland. Rangeland is not suitable for conventional crop production because of rainfall, temperature, soil conditions, slope, or other limitations. There are extensive areas in the western half of the United States that are designated as rangelands. Much of this land is in public ownership, however, many of the more desirable portions are in private ownership. This pattern of interspersed public and private ownership reflects the result of land claims being filed for valley sites that were suitable for homesteading. The higher elevation, more rugged lands remote from water were not deemed suitable for farming and remained in government ownership. Today, these publicly owned lands are managed by a variety of state and federal agencies, including the Bureau of Land Management in the Department of the Interior and the Forest Service within the Department of Agriculture. Indian tribes also control large areas of rangeland. State ownership is much less common than federal ownership but is significant.

In the western states that have large areas classified as rangelands, there is a system in place that allows ranchers to pay a fee for the right to run cattle on these lands. The rancher generally has a long-term agreement with the land management agency that provides sufficient permanence to allow the rancher to improve the land with fencing and water development so cattle herds can be maintained. Cattle numbers and grazing periods are specified in the grazing agreements.

Cattle or sheep grazing provides a way to harvest the forage that grows on these lands that would not otherwise be harvested. This forage is generally highly dispersed and may be available for only a few months. A typical scheme is for a rancher to own land in a valley where he or she pastures cows during the winter. The cows are bred to calf in the early spring when the weather is mild and the animals are still in pasture. As soon as the calves are old enough and the snow is largely melted from the higher elevation rangeland, and grass has grown sufficiently to support grazing, the cows with calves are moved to the rangelands. The animals remain on the range until fall. While the animals are on range, the rancher grows grass hay on the valley pastures. Many of the pastures are irrigated and fertilized to achieve maximum production. The hay crop is harvested two to four times during the summer and this hay is either sold as a cash crop or stockpiled for feeding the following winter. This latter decision is dependent on the relative number of cattle and the productivity and area of valley grassland. In the fall before the weather becomes severe, the cows and their calves, now several months old, are returned to the valley pastureland. The calves are branded, some are sold, and some that are now approximately 18 months old are transferred to feedlots for finishing. The winter pasture may be snow-covered or in a dormant stage so the hay harvested during the summer is hauled to the animals twice daily.

Impacts Water Quality

Prior to 1960, little attention was paid to the environmental impact of this ranching practice. More recently, however, it has been recognized that if the rangeland system is abused, significant environmental damage results. For example, if the range is overgrazed, weeds and less desirable grasses replace the native grasses and erosion losses increase. Because the streams that arise in these rangelands are the same ones that nurture the trout, salmon, and other game fishes, the increase in sediment loss tends to cover the gravel layers where the salmonids normally spawn, lay their eggs, and the juveniles are raised. Thus, overgrazing and improperly managed grazing are linked to decreases in sport fisheries.

Another concern of cattle raising on rangelands is related to damage to the streambanks and the riparian zone along the stream. Riparian areas that are in immediate contact with the stream and without animal impact tend to be areas of lush vegetation. This vegetation is not only conducive to animal grazing and loitering but also to maintaining the quality of the stream. It protects the streambank from erosion and acts as a vegetative filter removing particulate matter from runoff, absorbing water-carried nutrients, and providing shade to the stream.

Excessive cattle activity in the riparian zone can cause damage to the vegetation either due to removal or to compaction of the soil. Animal traffic also tends to trample the streambanks, promoting widening of the stream. Widening of the stream coupled with a decrease in shading vegetation has been linked with stream temperature increases. Although this process is still under study, cattle ranchers are responding with alternate management strategies to reduce the impact of cattle on riparian zones and streambanks. Cattle tend to congregate in the shady riparian zones during hot weather and to browse the lush vegetation. The riparian zone also may provide shelter from the wind during cold weather. One option is to fence the animals out of the riparian zone. This step is considered to be the most effective way to control cattle damage but is also an expensive step if the rancher has several miles of stream. An alternative is to lure the animals away from the stream by supplying water at a location convenient to the grazing cattle but away from the stream. Providing shade and shelter away from the stream is another helpful step. Controlling cattle numbers and grazing intervals also can prove effective. Figure 5.1 shows several best management practices (BMPs) integrated into a watershed with extensive cattle grazing.

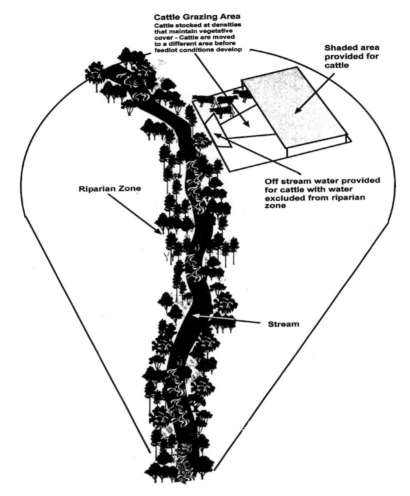

Cattle Grazing Area
Cattle stocked at densities
that maintain vegetative
cover - Cattle are moved
to a different area before
feedlot conditions develop

**Shaded area
provided for
cattle**

Riparian Zone

**Off stream water provided
for cattle with water
excluded from riparian
zone**

Stream

Figure 5.1.   Cattle grazing can occur within a watershed without damage to the water quality if an appropriate set of BMPs are incorporated into the pasture or rangeland management scheme.

There is also a concern with the immediate impact of cattle manure that is deposited in and immediately adjacent to streams. Because cattle drink from a stream and loiter in the riparian zone, they also tend to defecate and urinate there. Based on the observation that grazing cattle tend to defecate approximately 12 times a day and drop 2 to 3 liters of fecal matter each time, the load of indicator bacteria, fecal coliform (FC), and fecal streptococcus (FS), is significant. This is of particular concern where rangeland grazing areas are immediately upstream of recreational areas. Rangeland streams immediately below popular cattle watering areas frequently have FC and FS concentrations in excess of the standard for recreational waters where contact is involved.

Concern also has been expressed about the impact of runoff from rangeland grazing areas on other water quality parameters. Data tend to show that if the range is maintained in good condition, the vegetation is intact, and animal numbers are appropriate to maintaining the vegetation, runoff does not carry sufficient biochemical oxygen demand (BOD), nutrients, or indicator organisms to the stream to provide a measurable impact. If the range has been damaged, however, erosion losses carry large volumes of sediment that are damaging. Thus, rangeland grazing, although a long and well-established practice, has the potential for significant water quality impacts. These are nonpoint sources of pollution, hence they are difficult to document in conventional terms, but they are of importance as state agencies work to restore water quality and fisheries. Some of the measures that can be adopted to minimize the impact of grazing on rangelands are listed in Table 5.1, however, rangelands are not uniform, hence the greatest pollution control measure is a well informed and environmentally committed rangeland manager and rancher. Good ranchers have intimate knowledge of the areas for which they are stewards and can be highly effective in placing cattle when and where their grazing does minimal damage.

Other Environmental Quality Impacts

There is a romantic notion associated with cattle ranching that is difficult to associate with the environment. Ranchers typically live in remote areas and learn to rely on their own skills for most of the repair and service needs. They also have a long tradition of being persons closely in touch with their environment. Many people support the concept that rangeland well managed for cattle or sheep grazing is better protected than if the grazing were eliminated. For these policy issues, there is no easily identified answer.

Cattle feeding on rangelands is generally conducted in association with the native game of the area. One thought is that grazing promotes the population of elk, moose, and deer of the area by providing a supplemental food source. Regardless, livestock grazing is a long-established tradition on a major portion of the land in the western United States. It is important that grazing be conducted so as to preserve the quality of the resource and to do so in a way that responds to the needs of a large group of people who have traditionally been free of detailed regulations and administrative oversight.

Table 5.1.   Management measures that have been demonstrated to reduce the water quality impact of cattle grazing on rangelands

Measure	Impact	Difficulty
Construct fences along the upper edge of the riparian zone to exclude cattle from this part of the pasture.	Prevents riparian zone damage due to cattle and allows damaged riparian zones to recover.	Fence construction and maintenance are expensive in the quantity required. Exclusion from the stream requires an alternate water supply.
Provide an alternate water source and shade so that cattle do not need to go to the stream.	Reduces the amount of animal activity in the stream and riparian zone by 80 to 90%.	Less expensive than fencing.
Manage cattle numbers and grazing times to harvest the forage but not allow overgrazing.	Current state of management on most ranches.	Depends upon the judgement of the local land manager and is difficult to monitor and demonstrate improvement
Provide engineered watering access points along the stream that allow cattle to drink but prevent them from depositing manure in the stream.	Reduces the concentration of indicator organisms to downstream recreational water users.	Structures tend to be difficult to maintain, particularly during high-flow conditions. No more effective than an off stream water supply.
Provide an alternate water source along with divider fences that allow various portions of the range, including the riparian zone, to be grazed at the appropriate times and intensities for greatest range health.	Provides the greatest protection to the range but also places the management responsibility on the land manager.	This practice would be the one supported by professional ranchers, but it is difficult to monitor and ensure the lay public that it is being effectively pursued.

Pasture Production

Pasture rearing systems are popular throughout the world for a variety of animals. Pastures may be open public land with little or no management of the vegetation or highly manicured and elegantly fenced paddocks for confining prized racehorses. In between these extremes are pastures used for raising cows and calves, feeding and resting dairy cattle, and maintaining horses. Sheep, goats, and pigs also may be raised under pasture conditions.

The essential criterion of a pasture system is that vegetation is maintained throughout the area of the pasture. Pastures that are stocked with animals at too high of a rate develop bare areas that begin to operate more like feedlots and hence must be managed as feedlots with runoff control.

Stocking density and grazing duration are the most important variables in managing a pasture operation. In addition, the amount of feeding done by the animals impacts the carrying capacity of the pasture. Pasture development or improvement is another way to increase the animal carrying capacity. Pastures may be seeded to more desirable grasses and may be irrigated to promote improved feed production and improved vegetative cover. Grazing management is also important. For optimal production and animal performance, pasture areas are frequently allowed to rest without grazing during reseeding times and when new, fragile vegetation is being established.

Pastures for Dairy Cows

Pastures are important components for dairy production. Many small dairy operations in warm climates use pasture as their sole housing. Farmers maintaining fewer than 10 dairy cows for family or neighborhood milk supplies also may keep their animals on pasture. Larger dairies in the northern states are likely to use pastures during the summer to supplement stored or purchased feed. Grazing is particularly beneficial as a way to harvest forage from areas that are not suitable for mechanical harvesting. One of the labor inputs of using pastures for dairy cows involves bringing the cows into the milking parlor or barn twice a day for milking. Young calves may not be put on pasture during the time they require more detailed watching. Bulls may be inappropriate for pastures because of their tendency to damage fences. The appropriate stocking density and grazing interval are very local decisions. Stocking densities range from 5 to fewer than 0.05 animals per acre. The advantages and disadvantages of using pastures as part of a dairy are summarized in Table 5.2.

There are alternate ways to use pasture areas. The simplest is continuous pasture in which the land is always in grass. Continuous pasture is usually reserved for parcels of land that are not suitable for crop production. Pastures also may be part of a crop rotation system in which a particular parcel is in pasture for a while and after a year or two is planted to crops and another area used as pasture. There are also temporary pastures on which a forage material is available for a short period. For example, turning the dairy cows into a cornfield after harvest to glean the ears of corn missed by the picker, either mechanical or human. There are also seasonal pastures. A certain area may be set aside for winter grazing as part of a range system. In addition to these classifications, there are alternate ways to manage pasture areas. Pasture management strategies are summarized in Table 5.3

Table 5.2.   Advantages and disadvantages of including pasture feeding as part of a dairy operation

Advantages	Disadvantages
Less prone to disease spread than total confinement	Greater labor moving cows to and from pastures
Lower building construction costs	Cold weather pasturing increases maintenance ration requirements
Less management skill required than total confinement	Does not meet the feed requirements of high-producing cows; additional concentrate feed needed
Improved fertility and reproduction among the cows	Greater loss of forage from trampling and manure fouling than if cut, cured, and fed in confinement
Lower feed costs than purchase of all feeds	Milk production may fluctuate in response to pasture quality and weather

Pastures for Beef Cattle

Most of the considerations for dairy cattle also apply to pasture operations for beef cattle. There are a lot of beef cattle raised in pastures, particularly on small acreages by part-time farmers. In general, beef cattle on pasture gain at a lower rate than those fed a high-energy ration in confinement and may produce a slightly different quality of meat. Some consumers prefer a grass-fed beef, others the feedlot finished animal. Pastures for beef cattle may be irrigated for greater production.

Water Quality Impacts

Pasture production has historically been considered among the more pastoral agricultural operations and was not associated with water quality concerns. More recently, as greater attention has focused on nonpoint source pollution, the pollution potential of pasture-raised animals has become more evident and additional attention has been devoted to minimizing such impacts. Pasture production has water pollution pathways similar to those described for range production, but because pasture production systems are generally more densely stocked than ranges, the possibility exists for more intense pollution.

Two situations lead to water pollution from pastures. There are the ongoing or continuing sources such as animals defecating directly into or immediately adjacent to streams or lakes, and overflow from watering devices that flows to a water body. There are also runoff-related pollution sources. Runoff-related water quality degradation generally arises from sites associated with the pasture where vegetation has been removed or trampled to the extent that the natural filtering effect of vegetation is no longer effective. Thus, the most important pollution avoidance measure

Table 5.3. Alternate pasture management strategies

Strategy	Description	Advantages and disadvantages
Continuous grazing	Cows are allowed to graze the entire pasture for the full grazing season.	Simplest management but leads to inefficiency due to trampling. Cows may preferentially leave certain forage species.
Rotational grazing	Pasture is divided into two or more segments. A segment is grazed until the forage is exhausted and then the cows are moved to another segment and the first segment allowed to recover.	Pastures remain healthier and grow fastest during the resting periods. More fencing is required. Water and often shade must be provided in each of the segments.
Selective grazing	Producing cows are turned into a pasture segment and allowed to graze the best-quality forage. Then the cows are moved to another rested segment and dry cows and heifers moved in to eat the remaining forage.	Provides the best-quality forage to the cows that can best use it. Requires astute management to move cows at the appropriate times.
Strip grazing	This is the extreme of rotational grazing in which temporary fences are used so the cows are allowed to graze a fresh strip of the pasture each day. Frequently, a common watering and shade area serves all the strips.	This system is more labor-intensive because the fence must be moved frequently.
Zero grazing	The cows are not allowed to graze. The forage is harvested as green cut and delivered to the cows in confinement.	This system harvests the maximum amount of forage but requires more labor and equipment. Less forage is wasted and the cows are available nearby for milking.

in a pasture is the maintenance of a productive vegetative cover throughout the pasture. Maintaining this quality of cover is difficult and in any particular pasture, there may be sites where other remedial measures are required. For example, if an off-stream watering source is established to prevent the animals from having stream or lake access, the area immediately adjacent to this water source is likely to have sufficient animal traffic such that vegetation cannot be sustained. Typical measures to prevent water pollution from pasture operations are listed in Table 5.4.

Table 5.4.   Measures to prevent water pollution from pasture areas

Measure	Effect	Logic
Maintain a vegetative cover throughout the pasture.	Vegetation effectively filters particulate matter from runoff carried manure particles.	The surface of the pasture can act as a water treatment surface if the vegetation is in place. Bacteria are filtered from the overland flow and the vegetation uses the trapped nutrients. Depending upon local weather and soil conditions, specialized pasture management may be required. These measures may include irrigation, fertilization, weed control, and resting to allow the vegetation to recover.
Locate high-traffic areas away from any stream or standing water.	High-traffic areas are likely to become compacted and subject to increased runoff. These areas also are prone to becoming muddy.	By having these sacrifice areas away from sensitive waters, sufficient vegetated areas downslope can treat the runoff similarly to a grassed waterway, preventing manure-laden water from damaging the protected waters.
Prevent leaking or overflowing watering devices.	Avoiding continuous water sources within the pasture prevents muddy spots with their loss of vegetation and increased runoff.	Wet manure and saturated soils are more prone to overland transport than are dry materials. Saturated soils do not support vegetation.
Provide hard surfacing in intense-use areas such as adjacent to watering devices.	By providing a concrete pad or other surfaces adjacent to watering devices, muddy conditions can be avoided.	Depending on the number of animals in the pasture it may be impossible to maintain vegetation adjacent to watering devices. If this is the case, surfacing increases animal comfort and reduces the runoff of manure.
Ensure there is a well vegetated path for runoff from surfaced areas before it enters protected waters.	Vegetated areas effectively filter and treat contaminated runoff from hard-surfaced areas, reducing the solids, bacteria, and nutrients carried to the waterway.	Grassed waterways are low-cost effective measures to deal with small quantities of contaminated runoff.

Other Environmental Quality Impacts

Pasture production schemes are generally considered to be free of most environmental quality impacts other than water quality. Although there may be occasional odors detectible from pastures, these are usually of an intensity and quality that is found acceptable. As long as the pasture is managed to preserve vegetative cover, dust problems are avoided. Similarly, noise is generally not considered a problem in pasture operations.

Open Lot Confinement Systems: Feedlots

Open lot confinement is a popular way to produce a large variety of livestock and poultry throughout the world. Systems range from a simple pen confining a single animal to be fed, watered, and monitored to a highly sophisticated, concrete-surfaced feeding pen for dairy or beef cattle that are provided frequent feed, continuous water, and some degree of weather protection. The defining condition is that the animal or animals are confined at a sufficiently high density that vegetation no longer covers the surface. Manure collects on the surface and during periods of rainfall, a portion is transported off the surface. In Chapter 4, the quantity and quality of this runoff were described in sufficient detail to justify that this is an important waste source, which if not adequately managed has the potential to cause significant damage to regional streams and waterways. Runoff from feedlots was identified as a critical water quality issue in the 1972 Clean Water Act in response to a number of fish kills that had occurred. Thus, considerable thought and attention have been given to this livestock production scheme and pollution abatement measures are available to prevent water pollution. There are also air quality impacts of feedlots that are not as routinely solved as water pollution. This chapter describes how to design outdoor, unroofed confinement areas to minimize adverse environmental impacts.

Feedlots represent what is typically called confinement livestock production. Both range and pasture production depends upon large areas to assimilate manure. In confinement systems, manure is deposited in a much smaller area, hence provision must be included for its removal. Along with this additional design need, there are other implications of confinement that must be considered. Insects, especially flies, are attracted to manure. An effective feedlot design considers the control of flies. Similarly, dust is an attribute during the dry season in many feeding areas. Odors are an increasing concern related to outdoor confinement areas. The design of the facility impacts the extent to which odors are produced. Site selection, an important aspect of all confined animal feeding operations, is of importance relative to feedlots. Feedlots located inappropriately with regard to housing, commercial or recreational areas, schools, or churches are likely to encounter problems.

Cattle Feedlots

Most beef cattle destined for the food market are finished on a feedlot in the United States. They are likely to enter the feedlot at approximately 600 pounds and remain there, receiving a high-energy ration, for approximately 6 months until they achieve a market weight of between 1,100 and 1,200 pounds. The size of the feedlot may range from a few hundred animals up to 100,000 animals. The design of the feedlot depends upon the location. In the southwestern United States, where rainfall is markedly less than annual evaporation, the animals most likely are confined on a compacted earthen base, whereas in the more humid areas, concrete surfacing is more common to avoid muddy conditions. Muddy lot conditions adversely impact the rate of gain of the animals, contribute to fly and odor problems, and make animal management more difficult. Figure 5.2 shows the schematic of a cattle feedlot runoff control system. Notice that clean water is diverted around the manure-covered areas and that all pen drainage and other waters that may have contacted manure are collected, stored, and applied to cropland according to agronomic application rates. Table 5.5 lists some of the factors considered in feedlot design along with typical values.

Open Lots for Dairy Cows

Open lots similar to those for beef cattle are used for dairy cattle in areas where minimal environmental protection is required. The southwestern United States has many large dairy herds in open lots. The lots are frequently paved or partially paved and provided with sunshades to allow the animals to escape the sun. Lots are typically designed to allow 100 to 300 square feet per cow, depending on the amount of paving. Shade requirements are highly local but 35 to 50 square feet per cow is typical. Lots are generally sloped to drain.

Dirt Lots for Other Species

Pigs also are finished on dirt lots, however, the state of the art in the swine industry over the past decade has been toward housed confinement. Open lots are typically labor-intensive and have animals that perform less efficiently and that are more prone to disease than confined animals. Although stocking densities vary widely and are locally determined based on weather and soil conditions, stocking rates of 50 to 200 head per acre are typical. Feed is often available continuously from self-feeders and water from an automated trough.

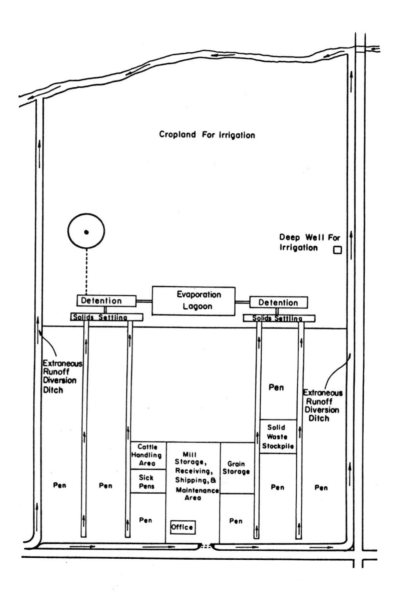

Figure 5.2. Schematic plan of a cattle feedlot showing the diversion of clean runoff around the manure-covered areas and the collection, pretreatment, and storage of manure-laden runoff.

Table 5.5.   Cattle feedlot design practice

Component	Typical values	Comments
Space per animal	400 sq. ft in unsurfaced lots without shelter 100 sq. ft in concrete-surfaced lots with no shelter 50 sq. ft in concrete-surfaced lots with shelter	Space requirements are based on the need to maintain lot conditions conducive to rapid gain. During the winter in humid conditions, large areas are required to avoid deep mud. During the dry season, crowding the animals helps control dust. Concrete lots prevent muddy conditions so greater densities are possible.
Feedbunk length per animal	2.0 ft twice daily feeding 0.5 ft feed continuously available	Feedbunks are placed adjacent to a service road to facilitate delivery in a feedtruck that disperses feed into the trough as it drives along the road. Concrete aprons are typically provided adjacent to feedbunks in unsurfaced feedlots to prevent muddy conditions in this high-traffic, high manure discharge area.
Watering facility	One space per 40 head	Watering tanks are frequently placed in the center of pens or along the dividing fences. Concrete aprons are frequently installed adjacent to watering tanks to avoid muddy conditions in this high-use area.
Slope	Desirable 4% Acceptable 2–8%	Feedlots exist with almost no slope to more than 15%. Flat lots tend to accumulate more water, dry more slowly, and develop low spots that collect water. Steep lots cause more manure to be carried away in runoff
Pen size	100–400 animals	Pen size is a function of the site but is generally in this range to facilitate inspecting and cattle movement.
Dust control facilities	Overhead sprinklers placed along pen partitions to maintain a sufficient surface moisture content to control dust.	Dust control is particularly important in the more arid areas. Excessive dust makes feedlots more difficult to operate, offends neighbors, and contributes to respiratory problems among the cattle.

Table 5.5.   *(continued)*

Component	Typical values	Comments
Frequency of cleaning	Once or twice per year	Cattle feedlots are cleaned to remove accumulated manure and restore the pen surface. Most lots are cleaned to leave the compacted base in place because this layer effectively prevents downward movement of soluble material. Manure may be mounded between pen cleaning to provide an elevated resting area for the cattle and an improved footing in the other areas.
Fly control	Combination of manure management, drainage, and chemical control.	Flies are a nuisance around feedlots and a frequent source of complaints. The most effective control is a well-designed lot that prevents wet areas, feedbunks that do not accumulate spoiled feed, and fence lines that can be kept clean. In addition, most cattle feedlot operators practice some chemical control.

Examples can be cited in which other animals are confined to unroofed feeding areas for a portion of their life cycle. In the past, large numbers of turkeys were raised in outdoor pens. More recently, there has been a trend toward housed systems for turkeys and other poultry species. For any species, controlling pollution from outdoor confinement areas includes the same essential features.

Environmental Quality Impacts

Runoff of precipitation falling on unroofed feeding areas is the most dramatic but not the only environmental impact to be managed. The material removed from the pens must be considered; the storage and ultimate disposal of this mixture of manure and soil requires a carefully designed strategy. Dust, odor, and fly control are concerns in any feedlot design. These issues are considered with cattle feedlots used as the basis for planning, but with the understanding that similar processes can be used for other species raised under similar conditions. An overriding variable in all of these situations is the weather in the area of the feedlot and the land uses in the proximity that could be potentially impacted. The distance from a feedlot that can be impacted by dust and odor is not clearly definable. There are examples of feedlots receiving complaints of dust and odors from property owners as far as 5 miles away; however, most concerns have been raised by property owners within a mile of the feed-lot. The size of the feedlot is an important variable in determining how far away the impact may be detectible.

Controlling Runoff from Feedlots

Most states have a designated water quality management agency respon-sible for the permitting of feedlots that confine 1,000 or more head of beef cattle, 700 or more dairy cows, and an equivalent number of other species. Some states have adopted programs that require permits for even smaller operations. It is advisable to be in contact with the appro-priate permitting agency in advance of any feedlot planning to deter-mine the detailed permit procurement process and to gain the benefit of any available local pollution prevention knowledge. The design of the feedlot provides several opportunities to minimize the amount of runoff. By taking advantage of these opportunities, considerable savings in con-struction and operating costs can be achieved. The overall strategy and a number of options for the control of feedlot runoff are outlined in Table 5.6. As this table indicates, there are several alternatives that can be se-lected depending upon local weather conditions and management preferences.

Feedlot design decisions to minimize runoff quantity and concentration. In the design of a feedlot, it is important to consider runoff control and begin by incorporating design details that may contribute to lower runoff vol-umes and to less concentrated runoff to be handled. Odor and fly con-trol programs also can be greatly facilitated by proper design of the feed-lot. The first principle of runoff control is to minimize the amount of runoff that is generated. Thus, it is important to hydrologically isolate the feedlot from the surrounding land so the runoff from nearby land that is not contaminated by manure is diverted around rather than through the feedlot. Land shaping and elevated berms can be con-structed to divert runoff from adjacent fields around the feedlot, assum-ing the feedlot is placed near the upper end of the slope. Good design insists that the site for the feeding pens not include a stream, river, or nat-ural draw that carries water after a rainfall event. A site should be selected with appropriate natural slope to minimize construction costs and facili-tate drainage. South-facing slopes tend to dry more quickly than other slopes and to have snow and ice melt sooner during winter feeding con-ditions. Slopes of 4% are considered optimal by most designers, however, others prefer to have the actual feeding pens designed at a lower slope, 2%, to reduce the amount of solid material transported off the pen surface.

Table 5.6. Options for the minimization, collection, treatment, storage, and land application of runoff from feedlots and other manure-covered surfaces.

I. Minimize the amount and concentration of runoff generated
 A. Locate the feedlot so runoff from adjacent areas does not flow over the feedlot
 B. Design the feedlot with appropriate slope and exposure
 C. Divert roof drainage and other uncontaminated water through underground piping
 D. Construct pens no larger than needed for the projected number of animals
 E. Install and maintain watering devices that do not leak and create wet areas
 F. Maintain feedlot surfaces free of wet or ponded areas
 G. Design and operate dust control sprinklers to avoid creating wet areas

II. Collect runoff in channels outside the pens
 A. Concrete lining is appropriate where roads cross channels
 B. Channels should be designed to flow greater than 2 ft per second to keep solids in suspension, but less than 5 ft per second to avoid erosion within the channel
 C. Lower velocity settling channel sections may be built to remove solids

III. Capture solids
 A. Low velocity (1 ft per second) channel sections
 B. Settling basins
 C. Grassed waterways

IV. Store in a runoff retention basin
 A. Single storm basin
 B. Extended storage capacity to provide flexibility

V. Apply stored runoff to crop or pastureland according to agronomic rates
 A. Sprinkler irrigation equipment
 B. Overland flow and infiltration systems

Pens should be designed to provide adequate space to handle the animals to be confined but excessive pen area leads to greater amounts of runoff to be managed. This option of reducing runoff quality has prompted some designers to move toward impervious surfaces rather than native soil so lots can be made smaller without encountering mud problems. Mud problems are primarily a cold-weather concern when there is limited drying. Concrete aprons behind the feedbunks for up to 20 feet also are effective in avoiding mud in these areas. There also should be a paved area around the watering device to make it comfortable for the cattle to drink and to avoid a muddy area around the water tank. Wet areas in a feeding pen contribute to greater amounts of runoff and to more concentrated runoff in two ways. Wet, muddy areas have less water storage capacity than dry areas and the manure already being partially in solution is more easily moved by runoff.

In the layout of feeding pens, it is appropriate to incorporate drainage ways outside the pens that can carry runoff from the pens in an orderly fashion and not interfere with use of access roads for feeding during wet weather. One option for accomplishing this end is to have the feedlanes located on parallel, elevated ridges and the pens sloped at approximately 4% away from the road toward a cattle-handling lane between pens that also doubles as a runoff flow channel during precipitation or snowmelt. The design of channels for conducting runoff away from the feeding pens deserves special consideration. These channels should be designed to carry runoff to the runoff collection basin without backing water into the pens. For the best operation, it is desirable to design the drainage system so that runoff from one pen does not flow into another pen. Where slopes allow, it is also advisable to incorporate some type of solids settling function into the runoff collection system. Solids that can be settled en route to the runoff retention basin are less expensively collected than those that enter the collection or storage basin. In the design of these channels, it is also appropriate to consider how they operate when snowmelt is the transport rather than runoff from rainfall. Snowmelt runoff is typically more heavily solids-laden than that from rainfall. It also may occur over an extended period, thus it would be desirable to trap it en route to the retention basin rather than to have it enter the basin, filling the storage space with a difficult-to-remove solid material.

Although not precisely a runoff control measure, the selection of watering devices is important to feedlot runoff. Good design ensures that water is readily available to the cattle and that they can access it regularly. Keeping the feedlot surface in good condition requires that the watering device not overflow or leak onto the pen surface. As mentioned above, wet manure tends to be transported off a lot more quickly than dry manure. If continuously flowing watering devices may be required to ensure open water during winter, an overflow pipe needs to be incorporated at the time of construction to prevent water from flowing onto the pen surface. Table 5.7 shows a checklist of considerations to minimize the amount and concentration of feedlot runoff. These considerations can be applied to the design of unroofed feeding areas for other species as well as beef cattle.

Conveying runoff from the feedlot to the retention basin. Rainfall events are part of the normal operation of feedlots, hence the system needs to be designed so that normal feeding and animal care can continue during runoff. This is normally accomplished by having the feed mill, feed storage, and other works associated with the feedlot located on the highest ground. Assuming a feedlot layout similar to that in Figure 5.2, the runoff would leave the lot, enter the runoff conveyance channels, and flow toward the lower end and be conveyed to the retention basin. Conveyance channels are typically designed as earthen channels with a 3 horizontal to 1 vertical side slope to reduce maintenance costs and to facilitate cleaning with a front-end loader or other conventional earth-moving equipment. It is next necessary to calculate the maximum flow rate to be handled at critical points in the collection system. This is done by first

Table 5.7. Considerations to include in the design of a livestock feedlot to reduce the amount and concentration of runoff to be managed

Principle	Design response
Minimize the amount of water that becomes contaminated with manure.	Locate the feedlot so that runoff from upland areas does not drain through the site.
	Construct diversion dikes and trenches to divert runoff from adjacent uncontaminated areas away from the feedlot.
	Construct appropriately sized pens for the number of anticipated animals. Avoid pens that are overly large for the length of feedbunk available.
	Collect any roof drainage or waterer overflow into an underground pipe and discharge it outside the feedlot runoff system.
Maintain the feedlot surface free of wet areas that would promote rapid runoff.	Use impervious access pads around watering devices to avoid creation of wet areas.
	Select watering devices that are less likely to overflow and leak onto the pens.
	Construct pens of appropriate slope and maintain that slope by frequent pen dressing, manure mounding, or other management.
	Design and operate dust control sprinklers to achieve uniform water distribution and avoid overapplication or runoff into low spots.
Trap solids carried in the runoff before they reach the retention basin.	Design runoff collection channels to have solid settling features at convenient locations where the settled solids can be collected.
	Avoid pen slopes in excess of 8% that tend to promote solids escape with runoff.

calculating the area from which runoff is being contributed at the lower end of the subchannels.

In the example calculations below, assume each of the collection channels serves 10 acres of feedlot. A typical design would be based on the maximum 10-year, 5-minute rainfall intensity. Data of this type are available from your local climatology department or more generalized summaries are published in the Livestock Waste Facilities Handbook, Midwest Plan Service. For our example, we select a 7-inch-per-hour rainfall intensity.

Selecting a channel slope and width involves the following considerations. Because 70 cubic feet per second is the maximum flow rate to be anticipated in this channel, it would be reasonable to allow a velocity somewhat in excess of what would typically be selected. By consulting the appropriate table (Tables 5.8–5.10), a collecting channel of 8-foot bottom width and 1% slope could be selected. This channel would accommodate a flow rate of 80 cubic feet per second at a depth of 1.5 feet and a velocity of 3.8 feet per second.

10 acres $\times$ (43,560 square feet/acre) = 435,560 square feet of contributing area

7 inches/hour $\div$ 12 inches/foot = 0.583 feet/hour

Calculating runoff rate in cubic feet per second

(0.583/60 seconds/minute) (60 minutes/hour) = 1.62×10^{-4} feet/second

(1.62×10^{-4} feet/second) (435,560 square feet) = 70 cubic feet per second

Solids removal from runoff. Once the material has been conveyed to the bottom of the feedlot area, a wider channel at less slope would be selected to allow some of the particulate matter to settle before reaching the retention basin. A channel with a bottom width of 20 feet and a slope of 0.1%, flowing 1.5 feet in depth, would have a velocity of 1.3 feet per second and allow solids to settle (Table 5.8).

Alternatively, it would be possible to select a holding basin where settling could occur. Such a basin would be shallow and have a gravity drain. In this case, it would be reasonable to select a basin with capacity for approximately 2 inches of runoff (one-third of an hour at the maximum anticipated rainfall rate). Figure 5.3 shows a typical settling basin that might be used in a cattle feedlot design. Such a settling basin might have a volume calculated as follows: (10 acres of feedlot) (2 inches per 12 inches per foot) = 1.6 acre-feet.

Thus, a shallow basin 3 feet in depth with a nominal surface area of 0.6 acre would provide time for the solids to settle. Such a basin should be designed to slowly drain into the runoff retention basin and if excess flow were to occur, it would immediately flow into the retention basin. Thus, the runoff retention basin would be protected from filling with solids and at the same time there was provision for collecting the total runoff up to the 24-hour, 25-year storm. An alternative is to construct a concrete-walled settling tank with a controlled drainage device to allow all of the early runoff to be collected. The early runoff carries a greater concentration of suspended solids than does runoff later in the storm. Earthen settling tanks also can be constructed and should be designed according to the same logic.

Table 5.8. Flow rate and velocity calculated for trapezoidal earthen channels at 0.1, 0.2, and 0.3% slope and various bottom widths. Side slopes: 3 horizontal to 1 vertical

Depth, ft	Slope,%	Unit of measure	Bottom width, ft					
			10	15	20	25	30	35
0.5	0.1	fps	0.6	0.7	0.7	0.7	0.7	0.7
		cfs	3.7	5.5	7.4	9.2	11.1	12.9
1.0	0.1	fps	0.9	1.0	1.0	1.0	1.1	1.1
		cfs	12.2	17.9	23.6	29	35	41
1.5	0.1	fps	1.2	1.2	1.3	1.3	1.3	1.4
		cfs	12.2	36	47	58	70	81
2.0	0.1	fps	1.3	1.4	1.5	1.5	1.6	1.6
		cfs	43	60	78	96	114	132
0.5	0.2	fps	0.9	0.9	1.0	1.0	1.0	1.0
		cfs	5.2	7.8	10.4	13.0	15.6	18.2
1.0	0.2	fps	1.3	1.4	1.5	1.5	1.5	1.5
		cfs	17.3	25.3	33	42	50	58
0.5	0.3	fps	1.1	1.2	1.2	1.2	1.2	1.2
		cfs	6.4	9.6	12.8	15.9	19.1	22.3

Source: Midwest Plan Service, Livestock Facilities Handbook, MWPS-18, 1985.

Table 5.9. Flow rates and velocity calculated for trapezoidal earthen channels at 1% slope and various bottom widths. Side slopes: 3 horizontal to 1 vertical

Depth, ft	Unit of measure	Width, ft						
		1	2	4	6	8	10	12
0.5	fps	1.4	1.8	1.9	2.0	2.1	2.1	2.1
	cfs	1.4	3.5	5.7	8.0	10.8	12.6	14.9
1.0	fps	2.3	2.6	2.8	2.9	3.0	3.1	3.2
	cfs	9.2	15.6	22.4	29.4	36.5	43.7	50.9
1.5	fps	3.0	3.3	3.5	3.7	3.8	3.9	4.0
	cfs	27.1	39.6	52.7	66.1	79.8	93.7	107.7
2.0	fps	3.7	3.9	4.1	4.3	4.5	4.6	4.7
	cfs	58.4	78.5	99.4	120.8	143	165	187
2.5	fps	4.2	4.5	4.7	4.9	5.0	5.2	5.3
	cfs	106	135	165	196	227	259	291
3.0	fps	4.8	5.0	5.3	5.4	5.6	5.7	5.8
	cfs	173	212	252	293	335	378	421

Source: Midwest Plan Service, Livestock Facilities Handbook, MWPS-18, 1985.

Runoff retention basin volume. The runoff retention basin is perhaps the most critical component of a feedlot runoff control system and the one that requires the most careful consideration regarding capacity. The basic guideline according to the federal permitting system is that the system cannot discharge except in a 24-hour, 25-year storm. By consulting regional climatic data, it is possible to determine the size of this storm

Table 5.10.   Flow rates and velocity calculated for trapezoidal earthen channels at 2% slope and various bottom widths. Side slopes: 3 horizontal to 1 vertical

Depth ft	Unit of measure	1	2	4	6	8	10	12
0.5	fps	2.0	2.5	2.7	2.8	2.9	3.0	3.0
	cfs	2.0	5.0	8.1	11.3	14.6	17.8	21.1
1.0	fps	3.2	3.7	4.0	4.2	4.3	4.4	4.5
	cfs	13.0	22.1	32	42	52	62	72
1.5	fps	4.3	4.7	5.0	5.2	5.4	5.5	5.6
	cfs	38	56	75	94	113	133	152
2.0	fps	5.2	5.6	5.9	6.1	6.3	6.5	6.6
	cfs	83	111	141	171	202	233	265
2.5	fps	6.0	6.4	6.7	6.9	7.1	7.3	7.5
	cfs	150	191	233	277	321	366	411
3.0	fps	6.8	7.1	7.4	7.7	7.9	8.1	8.3
	cfs	244	300	357	415	475	535	595

Source: Midwest Plan Service, Livestock Facilities Handbook, MWPS-18, 1985.

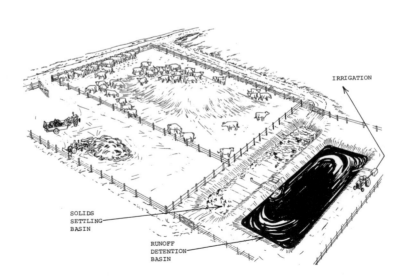

Figure 5.3.   Typical solids settling basin appropriate for inclusion in a feedlot runoff control system to prevent solids from accumulating in the runoff storage reservoir.

with relative precision. The state water quality regulatory agency or other state permitting authority also may have this information for various regions within its jurisdiction. An appropriate interpretation of this criterion is that at all times, there should be sufficient storage capacity available to accommodate the runoff from a storm of this size. Thus, the minimum size of a runoff retention basin would be the 24-hour, 25-year storm minus whatever volume would likely be held on the lot. This latter volume is often estimated as 0.5 inches for an earthen lot and slightly less for a concrete-surfaced lot.

Sizing a runoff retention based solely on the single design storm, in most cases, does not lead to the best design. It assumes that the basin is empty when the design storm occurs. In most areas, there is a period during which it is either not convenient or environmentally wise to pump from the runoff retention basin to cropland. Another consideration is that the runoff has agronomic value based on its nutrient content or on the value of the water when it is applied at an appropriate time. Thus, a designer needs to incorporate an additional management volume into the capacity of the runoff retention basin. Various researchers have studied this issue and created estimates of runoff retention basin volumes that accommodate more flexible management techniques (Table 5.11). Note also in Table 5.11 that the capacity of the pumping system that removes the accumulated runoff has a relatively small effect on the runoff retention basin capacity but a rather dramatic impact on the number of pumping days that is required on an annual basis. This factor may be important in the overall management of the feedlot and should be considered in the total system design. Note also that by increasing the runoff retention basin by approximately 25%, the operator gains significant additional flexibility in scheduling when to apply effluent to his or her land. If land availability is an issue, the larger volume becomes an additional benefit by allowing application after storage, which results in nitrogen reductions similar to those that occur in a lagoon.

The issue becomes how to size a runoff retention basin. This decision should be made with some understanding of the desires and management style of the intended manager as well as the local weather patterns. Several options for sizing the runoff retention basin are presented in Table 5.12. All of these options have been used successfully in some locations so none is clearly the best. Some, however, provide greater flexibility than others. State or local regulatory agencies also may have established minimum retention basin volumes. The important message herein is that the designer should not consider the minimum specified volume as necessarily the best option. Some serious thinking by the designer can make the life of the client much better for a very small cost difference.

Table 5.11.　Impact of management alternatives on the required runoff retention basin capacity (inches of runoff) and average number of days of pumping required per year

Policy		Design storm[1]	Pump rate[2] 0.25	Pump rate 1.0	Pump rate 2.0
Lubbock, TX		5.0			
Year-round disposal	Basin capacity		9.5	8.0	8.0
	Pumping days		21.8	4.3	1.6
Apply to corn plus a	Basin capacity		10.6	10.6	11.2
preplant irrigation	Pumping days		15.7	3.3	1.4
Apply to hay crop plus	Basin capacity		10.1	8.6	8.6
winter disposal	Pumping days		19.9	4.1	1.6
Pendleton, OR		1.5			
Year-round disposal	Basin capacity		2.0	2.5	
	Pumping days		4.6	1.0	
Apply to corn crop	Basin capacity		3.4	3.6	
	Pumping days		4.0	1.0	
Apply to hay crop plus	Basin capacity		2.0	2.5	
winter disposal	Pumping days		5.0	1.0	
Ames, IA		5.4			
Year-round disposal	Basin capacity		10.3	9.5	10.3
	Pumping days		44.0	10.3	5.0
Apply to corn crop	Basin capacity		16.4	16.6	17.3
	Pumping days		40.1	10.0	5.0
Apply to hay crop plus	Basin capacity		10.7	11.3	12.0
winter disposal	Pumping days		42.3	10.2	5.0

Source: Miner et al. 1979.
1. Design storm is the 25-year–24-hour storm.
2. Pumping rates expressed as acre in./acre of feedlot/day.

Runoff retention basin design details. Once a decision has been made as to the amount of runoff to be accommodated in a runoff retention basin, the rest of the design process is relatively straightforward. It is important to avoid the temptation to build a runoff retention basin by constructing a dam across a nearby draw, creating a pond or lake, and planning to use that as a runoff retention basin. Such a solution is doomed to failure because at the time of maximum feedlot runoff, the pond is receiving the maximum amount of runoff from its drainage area and the system discharge, creating a permit violation and potentially killing a large number of fish and damaging downstream water users. The runoff retention basin is an engineered structure with a dike constructed such that only runoff from the feedlot can enter. Typically, the system of runoff collection channels that serves the feedlot pens is extended to carry runoff from the solids settling area to the runoff retention basin. It is important that the runoff retention basin be impervious to water. A typical standard is a maximum infiltration rate of 10^{-7} centimeters per second. This

Table 5.12. Alternative strategies for determining the volume of a feedlot runoff retention basin

Option	Description
Winter runoff plus design storm	Provide sufficient volume to accommodate the runoff during the period for which application to crop or pastureland is inappropriate plus the runoff from the design storm and the volume that lands on the retention basin from the design storm
Twice-yearly application	Provide sufficient volume to accommodate the runoff during the wettest 6-month period over a 10-year period plus the runoff from the design storm and the volume that lands on the retention basin from the design storm
Annual application	Provide sufficient volume to accommodate the runoff for the wettest year in 10 plus the runoff from the design storm and the volume that lands on the retention basin from the design storm
Design storm plus 30 days	Provide sufficient volume to accommodate the runoff for the wettest 30-day period, 1 year in 10 plus the runoff from the design storm and the volume that lands on the retention basin from the design storm

means that for each acre of retention basin, less than 3,000 gallons seep through the bottom per month. For many soils, this degree of imperviousness can be achieved with appropriate moisture content and compaction. In other locations, it is necessary to import clay material to mix with native soil or to purchase an impervious liner that can be installed. Whichever approach is taken, it is important that the basin not leak and that testing be done that ensures this degree of protection.

Once the volume of the runoff retention basin has been calculated, the actual dimensions can be calculated. Some example volumes of rectangular earthen basins with sloping dikes are presented in Table 5.13. The volumes presented in this table are based on interior dike slopes of 1 foot of vertical rise for each 2.5 feet horizontal. Steeper slopes can be constructed in some areas, but as dikes become steeper, erosion control and dike maintenance become more difficult. Runoff retention basins are essentially regular trapezoids built into the soil. Although it is difficult to calculate the design dimensions from the volume, calculating the volume knowing the top dimensions, depth, and slope of the dikes is relatively easy. One way is to calculate the bottom area, top area, and the area at the mid-depth. If you multiply the mid-depth area by four, add the top area and the bottom area to that sum, and divide by six you have calculated a representative area that can be multiplied by the depth to reasonably estimate the volume of the basin.

Table 5.13.   Lengths and volumes of 20-foot deep earthen basins, of various widths, with interior dikes that have a slope 2.5 feet horizontal for each foot of vertical rise. Volumes are calculated assuming the top 2 feet are reserved as freeboard

Volume of liquid, thousands of cu ft	Interior width, 100 ft	Interior width, 150 ft	Interior width, 200 ft	Interior width, 300 ft	Interior width, 400 ft
	Interior length, ft	Interior length, ft	Interior length, ft	Interior length, ft	Interior length, ft
300	410	223	165	120	101
325	441	238	175	126	105
350	472	253	184	132	109
375	503	267	194	137	113
400	534	282	204	143	117
450	596	311	223	154	126
500	657	340	242	166	134
600	781	399	280	188	150
700	904	457	319	211	166
800	1,028	516	357	234	182
900	—	574	395	256	198
1,000	—	633	433	279	214

Source: Midwest Plan Service, Livestock Facilities Handbook, MWPS-18, 1985. Additional tables are available for alternate widths and interior dike slopes.

It is important for the owner to realize that a properly designed runoff retention basin may stand empty for several years in many parts of the country, particularly in areas with low rainfall but a pattern of intense stormy periods on an infrequent basis. The challenge is to continue to maintain "that empty hole in the ground" even though it has not had runoff in it for several years. The runoff collection channels need to be maintained and the retention basin dikes kept in proper repair.

Management of a runoff retention basin. The two major responsibilities of the feedlot manager regarding the runoff retention basin are to maintain the volume so there is always storage capacity available to hold the runoff from a design storm and to maintain dikes and channels so the system does not discharge due to structural failure. The first responsibility is typically met by application of the accumulated runoff to crop or pastureland with conventional irrigation equipment. The amount of runoff that is to be applied and the scheduling of the applications involve water and nutrient management and should be based upon crop requirements and the soil's ability to absorb water. These considerations are covered in Chapter 10 and are not discussed herein except to remind you that stored feedlot runoff is variable in content. Therefore, at least a partial nutrient analysis is generally required before each application season to adequately determine the amount to be applied. Table 5.14 was

Table 5.14.   Lengths and volumes of 10-foot deep earthen basins, of various widths, with interior dikes that have a slope 2.5 feet horizontal for each foot of vertical rise. Volumes are calculated assuming the top 2 feet are reserved as freeboard

Volume of liquid, thousands of cu ft	Interior width, 50 ft	Interior width, 75 ft	Interior width, 100 ft	Interior width, 150 ft	Interior width, 200 ft
10	86	55	46	—	—
20	148	83	64	50	—
30	211	110	82	60	—
50	336	166	117	81	66
70	461	221	153	102	81
100	648	305	207	133	103
120	773	360	242	154	117
140	898	416	278	175	132
160	—	471	314	196	147
180	—	527	350	216	162
200	—	583	385	237	176
250	—	721	475	289	213

Source: Midwest Plan Service, Livestock Facilities Handbook, MWPS-18, 1985. Additional tables are available for alternate widths and interior dike slopes.

compiled to indicate approximate concentrations of major nutrients in the material and the variability to be expected. The values are not, however, sufficiently well established to replace on-site analyses.

A comparison of the typical values in Table 5.15, quality of stored feedlot runoff, with the values for actual runoff given in Table 4.3, reveals a dramatic difference. This difference is important to the designer because it reveals that considerable nutrient loss takes place due to settling and volatilization. Nitrogen concentration has been reduced primarily through ammonia volatilization to the air. Phosphorus concentration has decreased through precipitation and settling to the bottom of the retention basin. Potassium, however, being the most soluble of these three nutrients, has changed relatively little. In the more arid portions of the country, feedlot managers pay particular attention to the total dissolved salt concentration to be certain they do not damage crops to which the material is applied due to excessive salt concentrations.

Water quality implications. This discussion supports the common interpretation that runoff from concentrated, unroofed livestock and poultry feeding areas is a high-strength organic waste material released in relatively large quantities over a brief period. If this material is not captured and managed, it has the potential to do extensive damage to streams, lakes, or other natural waters of the area. The impact of a feedlot area can be several times greater than the calculated daily manure production

Table 5.15. Characteristics of stored feedlot runoff at the time of application to crop or pastureland.

Constituent	Typical value	Range
Total solids, mg/l	2,000	1,000–5,000
Volatile solids, % total solids	50	40–70
Total nitrogen (N), mg/l as N	125	40–300
Ammonia N, % total N	40	30–60
Phosphorus (P), mg/l as P	30	20–100
Potassium (K), mg/l as K	250	200–700

because of this storage and sudden release principle. For feedlots with runoff capture and storage facilities, the risks are different. There are both surface and groundwater concerns that require consideration by the manager. If the runoff collection system is not maintained, runoff can escape due to eroded channels and bypass the retention basin. If the retention basin has not been maintained and the dikes are eroded, a more serious problem can occur. If a retention basin begins to discharge due to a damaged dike, the danger exists that this flow through a small opening in the dike may further erode the dike and ultimately result in a loss of the total contents. Although this latter disaster has proven infrequent, it is sufficient to prompt regular and diligent inspections. Another possibility is that the retention basin does not have adequate space to retain the runoff from a storm due to its not having been pumped at an appropriate time. This again is a matter of operator diligence and legal responsibility.

The more common source of surface water pollution from feedlots is overapplication or inappropriate application of solid manure or liquid to land or run off of this material due to continued application or a rainfall event immediately after application. In most areas, it is inappropriate to apply manure or accumulated runoff to steeply sloped land or to land immediately adjacent to a stream or other water body. The inclusion of a vegetated buffer strip between a stream or lake and any manure disposal area is an effective safety practice that reduces the likelihood of runoff causing damage. Similarly, manure and stored runoff should not be applied to land where it cannot infiltrate or be incorporated. Frozen, snow-covered, or saturated soils are inappropriate for manure application in most situations. There are people that argue that in some areas of flat fields, less than 1% slope manure can be applied to frozen or snow-covered land with minimal risk. If alternatives are available, these are risks that are best avoided.

There are also groundwater concerns associated with feedlots that deserve consideration. In the early days of feedlot operations, there was a concern that water landing on the feedlot surface would pass through the manure layer and transport soluble nutrients, especially nitrogen, to the groundwater. This fear is quickly dispelled by anyone who attempts to dig an observation hole through a feedlot surface. The action of animal hooves on moist manure and soil mixtures forms a high-density layer that is impervious to the passage of water. This same phenomenon was discovered by designers who installed drainage tile beneath a feedlot to overcome seasonal high water tables. The tiles remained dry but the feedlot surface was a wet, muddy mess because the water was unable to penetrate the compacted manure–soil interface. Thus, earlier concerns of infiltration through feedlot surfaces have been largely dispelled. There is a concern, however, if a feedlot is abandoned or if the decision is made to use a former feedlot area for crop production. Then, the landowner or manager needs to face the reality of a land area with a large accumulation of nutrients that require special management to prevent their escape.

Groundwater pollution from the runoff retention basin is a second concern. This concern is largely met by the designer ensuring that the permeability of the retention pond bottom and dikes is sufficiently low that leakage is within acceptable limits. The greatest actual danger is associated with the overapplication of accumulated runoff on crop or pastureland. Overapplication can occur due to failure to accurately calculate the amount of runoff water that can be safely applied to meet the nutrient uptake of the crop or to inattention to the equipment. It also can occur by allowing irrigation to continue beyond the planned interval or by overestimating the yield of the crop, hence its nutrient uptake, or by excessive rainfall or irrigation after runoff application moving the soluble nutrients beyond the root zone of the crop.

Odor concerns. Cattle feedlots have frequently spurred complaints of elevated odors from neighbors and from people driving past the area. Although it is difficult to envision a feedlot without any odor, it is certainly possible to have feedlots that are less odorous than many have been in the past. Although odors are discussed in detail in Chapter 12, it is appropriate to look at specific odor control possibilities of open feedlots while discussing feedlot design.

Odors come from a variety of sources around a cattle feedlot. Many ingredients in cattle feed have odors. Waste potatoes are a prime example. Potato waste is frequently hauled to a feedlot from a nearby potato processing plant and stored in a lined pit. This pit has an odor. Silage made from vegetable processing waste is another odorous material. Each of these odor sources deserves particular attention. Other odor sources include dead animals that are allowed to remain on site too long. The best solution to the dead animal odor problem is prompt disposal to a rendering operation, composting, or burial according to locally established regulations. The most common odor around a feedlot and the one that creates the most frequent complaints is that of anaerobically decomposing manure. Odors from decomposing manure can be reduced by avoiding an accumulation of moist manure in low spots and along fence lines of feedlots. Mounding of manure to improve drainage is helpful as is properly draining and rapidly drying pens. Feedlots with southerly slopes generally dry more quickly than those with northerly or easterly slopes. Specific feedlot design and operating details to reduce the frequency and intensity of odors are listed in Table 5.16.

Feedlot dust control. Feedlots pose a particular challenge in controlling dust. Dust is frequently a nuisance condition to neighboring residents as well as an economic issue to the feedlot operator because it interferes with the health and well being of the animals. Dust is most frequently a problem during the early evening hours (around sundown) when the animals are active and stir up dust on the feedlot surface. Feed trucks, horses, and general vehicle traffic around the facility also contribute to dust problems. Dust is most often reduced by increasing the moisture content of the feedlot surface layer. Sprinkling systems and mobile water trucks fitted with a spray nozzle are being used. Solid set sprinkler systems have the best potential for success but must be managed precisely to avoid overapplication of water to any particular area that is likely to lead to increased odor. No progress is made if a dust problem is traded for an odor problem. The pens need to be uniformly graded if sprinklers are to be effective. Lots with low spots that collect excess water are sure to create odor problems.

Fly control on feedlots. Flies like dust are both a nuisance and an economic concern around feedlots because they bother neighboring residents and interfere with animal performance. Most feedlot operators control flies by having a facility that was designed to minimize fly breeding opportunities and by operating the facility so there are as few opportunities as possible for flies to lay their eggs so the larvae hatch in moist manure. Because accumulated wet manure and spilled feeds are the prime breeding grounds for flies, odors, dust, and flies are interrelated. Practices that minimize odors also tend to reduce fly breeding. Dusty feed pens are generally too dry for fly breeding, however, it is possible to have dry conditions and dust in some parts of a lot and wet manure with flies and odor production in another. Lot shaping is important because it prevents the development of low spots that accumulate moist manure. Mounding also can contribute to a well-drained lot. Attention to feedbunk design so that

Table 5.16. Feedlot design and operating techniques to reduce odors

Odor source	Cause of the odor	Methods to reduce or minimize
Pen surfaces	Anaerobically decomposing manure either on the pen surface, along the fence lines, in low wet spots within the pen, or on the animals.	Reduce the moisture content of the manure, clean along the pen partitions, fill the low spots to improve pen drainage, lower the animal density, construct hard-surfaced areas around feedbunks and watering devices.
Feed storage areas	Odors are associated with certain feeds, especially those feeds derived from food processing wastes.	One option is to cover the feed storage device with an odor-capturing cover. Other possibilities include promoting a dry crust on top of the storage, reducing the surface area of the storage, or going to an enclosed storage system.
Dead animals	Decomposing dead animals are both an odor source and a visual affront to many people. In addition, they are a public health hazard. Most areas have specific regulations with regard to how long they may remain on site.	Prompt removal to a rendering plant is the most desirable alternative for the disposal of dead carcasses. A second alternative is composting as discussed in Chapter 9. A third alternative is burial. Specific guidelines are available with regard to burial in most areas to protect groundwater and to prevent the spread of disease due to scavenging animals. Carcasses also may be incinerated in appropriately designed and locally approved facilities. Open burning is not an acceptable alternative.
Application of stored runoff	Irrigation of stored feedlot runoff creates a large surface area and promotes the escape of volatile odorants.	If irrigation disposal is infrequent, it may be possible to schedule it during a time when odors are blown in a direction that creates fewer problems. Odors are also generally less of a problem if irrigation is done early in the day, during sunny weather, and when winds are blowing at more than 6 miles per hour. Lowering the pressure and increasing the droplet size also decrease the exposed surface area.
Overall feedlot area	Although odors are caused by definite sources around a feedlot, the impact on neighbors is more complex than the odor release and improving overall feedlot appearance may reduce odor complaints.	Appropriate visual screening, planting of trees and generally improving the visual appearance of a feedlot do much to improve the perception of the operation.

there are not piles of moist feed available for fly production is helpful. Manure that gets trampled regularly does not harbor a large fly population because of the physical damage by the animals' hooves. Having a tractor with a blade that can remove accumulated manure along the fence line is helpful because animal traffic tends to create manure ridges in areas that receive less animal impact.

Flushed and Other Dilute Manure Handling Systems

Hydraulic manure transport became popular shortly after it became fully appreciated that confinement production of animals larger than poultry could be economically moved into houses and that their feeding, watering, environmental control, and manure handling could be automated. Dilute manure handling systems have continued to evolve because more sophisticated valving has simplified the flushing process. The common idea in all flushing systems is that manure can accumulate in a gutter, alley, or shallow pit of some type and periodically, a large volume of water is flushed through this area, carrying the deposited manure out of the building to a treatment or storage area. The advantage of this system is that it allows for frequent manure removal, thereby potentially minimizing odors. It is also a process that can be automated so that it requires very little operator attention and produces a dilute manure that flows easily through pipes or other hydraulic handling equipment, assuming a nonfiberous feed ration. Two alternative gutter arrangements are shown in Figure 5.4. There are also building designs that incorporate flushed gutters within enclosed buildings.

The most obvious disadvantage of flushing manure systems is that they create a large volume of wastewater for storage or treatment. Many systems use up to 10 volumes of water for each volume of manure. In mild climates where manure can be land-applied throughout the year, this water may not be a particular problem. In colder climates or other situations in which there are extended periods in which fields or pastures are unavailable for wastewater disposal, this large water volume may make storages too large and expensive. Flushing systems are also popular where it is acceptable to partially treat manure and then discharge effluent to natural streams and waterways. Early confinement systems in Taiwan, Thailand, and the Philippines were designed to use large volumes of water that were then treated and discharged to nearby streams. This practice is attractive as long as the criteria for discharge can be met relatively easily but it becomes more expensive as discharge quality criteria become more restrictive.

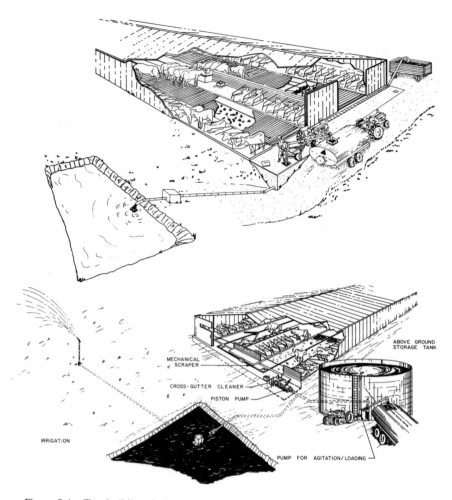

Figure 5.4. Two building designs incorporating gutters for the transport of manure from animals in confinement.

There is also the possibility of water reuse in flushed systems. Particularly in systems in which the animals are separated from the flushwater, it is possible to treat the water sufficiently for recycling. Concerns with this practice are potential odors within the building and the precipitation of magnesium ammonium phosphate (struvite) on the inside of recycle lines. Recycled water is commonly used to flush swine, poultry, dairy, and beef confinement facilities.

Gutter Design Considerations

Gutter slope, length, and flush volumes are related and must be considered in the design of a gutter that can effectively remove manure while not wasting water. Gutter slope is important because if the gutter is too flat, the flush does not have sufficient velocity to transport the manure. If the slope is too great, there is not sufficient depth of flow to lift the manure off the floor. In either case, manure is left in the gutter, contributing to odors within the building.

The rigorous analytical design of a flushing gutter is a complex problem because the system is continuously in an unsteady state. Most designers have, however, elected to use the traditional Manning's formula that is for the design of steady-state flow in channels, then to modify these results based on their personal experience and those of other designers. Manning's formula is as follows:

$$V = \frac{1.486\ R^{0.67}\ S^{0.5}}{n}$$

where V = velocity in feet per second; R = hydraulic radius of the channel, cross-sectional area divided by the wetted perimeter; S = slope of the channel; and n = the roughness coefficient. The roughness coefficient is typically 0.015 for concrete channels, a value of 0.020 to 0.030 is frequently used to account for the presence of manure in the channel.

This equation can be rearranged to calculate the slope directly if the water depth and velocity have been selected.

$$S = \frac{n\ V^2}{1.486\ R^{0.67}}$$

Consider an example calculation. A water depth of 2 inches (0.167 feet) and a velocity of 3 feet per second, n = 0.015, and a channel width of 3 feet would be reasonable values to insert into the equation. Thus,

$$R = \frac{Area}{Wetted\ perimeter} = \frac{3\ (0.167)}{3 + 2(0.167)} = 0.15$$

$$S = 0.012\ or\ 1.2\%$$

This suggests a slope of 1.2% would be appropriate. Gutters of 50 to 100 feet are typical. It is frequently desirable to slope gutters from each end toward the center if the building is more than 200 feet in length. This channel would carry water at a rate of 1.5 cubic feet per second. Assuming a typical flow duration of 10 seconds, the volume of water required for each flush would be 15 cubic feet or approximately 115 gallons.

Experience with channels sloped between 1 and 1.5% has generally been positive. One important consideration is that channel bottoms must be relatively smooth for flushing to be effective and the channel bottom needs to be constructed flat from side to side to avoid having the flow concentrated in one area and another area not being adequately flushed. If it is desirable to construct gutters no more than 6 feet in width beneath slotted floors, most designers elect to divide the channel into units of 4 to 6 feet in width to better control water distribution. An exception is alleys in free-stall dairy barns that may be 10 feet in width and that are frequently flushed as single gutters.

There are a variety of devices being used to produce the flush. The simplest is an elevated storage tank placed at the upper end of the gutter that is fitted with a timer-controlled, quick-opening, motor-operated valve. If this kind of valve is used, it is desirable to have an overflow pipe on the tank to prevent uncontrolled overflow. Another option is to flush by directly pumping to the upper end of the gutter with a high-volume pump or by the use of a tipping-bucket style of flushing device. Alternate flushing devices are shown schematically in Figure 5.5. Table 5.17 lists the more common flushing devices along with some of their advantages and disadvantages. The choice of flushing devices is often a matter of personal preference.

Flushing Systems for Pigs

In many parts of the world, flushing systems for pigs are popular. The most typical arrangement is the partially slotted floor design shown schematically in Figure 5.6. In this system, 15 to 25% of the floor area is slotted to allow manure to fall through the open areas to a shallow gutter. Animals may be fed at the upper end of the pen away from the slatted floor section with either self-feeders, or any of the various feeding systems. Watering devices are generally located over or near the slatted floor section so that any wasted water does not contribute to a wet pen and lure the animals to that end of the pen. The gutter is usually designed with a flat cross section and is sloped at approximately 1 to 1.5% in the direction of the length of the building. One to four times per day a flush of water, either fresh or recycled, is introduced at the upper end of the building and flows as a wave the length of the gutter, lifting and carrying the accumulated manure from the building. In parts of the world where pigs are washed frequently, such a system accommodates this practice. Most researchers seem to agree that the daily washing of pigs is not necessary for their performance. If cooling is needed, an intermittent spray near the slotted portion of the floor is more effective and results in less water being used and therefore less wastewater for handling.

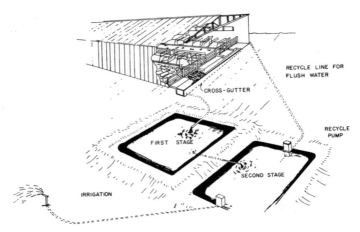

Figure 5.5.   Example of a flushed gutter system for transporting manure from a free-stall dairy barn. Similar gutters can be incorporated in building designs for other species.

Various alternatives to this basic design are in use. A less expensive system is the open gutter design (Figure 5.4), in which the gutter is part of the pen floor, typically 3 to 5 inches in depth, and accessible to the pigs. This design has been particularly popular for finishing pigs and for gestating sows. In some areas, however, developing recommendations to eliminate animal contact with recycled wastewater are reducing its use. Pens for farrowing buildings and nursery-size pigs are usually designed to achieve greater ensurance that the pens remain dry. Farrowing pens are often totally or partially slotted. There may be a solid floor area for baby pigs to be kept as warm as possible. Nursery pigs are likely to be raised on totally slotted floors from the time they are separated from the sow until they enter the finishing building. For small pigs, the floor may be elevated and made of open mesh to ensure they remain dry as a method of disease control.

There is a modification of the flush system for swine that is worthy of mention because it has the capability of getting the manure out of the confinement building at a concentration of closer to 2% rather than 1%, which is typical of flushing systems. It is called the pit precharge system and in this system, a flat, shallow pit with zero slope is constructed beneath the slotted floor area and is precharged with 3 to 5 inches of water. Every 3 to 7 days, the pit is quickly drained, creating sufficient water velocity to carry any solids from the bottom of the pit. This creates a water motion similar to flushing if properly designed. It is important in this design to create enough velocity in the water to remove solids and to drain

Table 5.17.   Alternate devices for flushing gutters

Device	Capacity, volume	Advantages	Disadvantages
Elevated tank with electrically controlled valve on the droppipe	100–200 gal	Simple concept Easily adjusted flushing interval	Automatic valves require frequent maintenance
Elevated tank with automatic siphon	100–200 gal	Low cost Tank automatically flushes when water volume reaches set point	Demands careful construction too achieve reliable flushing
Floor-level siphon tank	200–1,000 gal	Easier to build and larger capacity than elevated tank	Consumes floor space that could otherwise be pen space.
Drop side tanks	300–3,000 gal	Simple in concept Easily built on farm Large capacity	Drop door may be heavy Consumes space Difficult to achieve water tightness
High-volume pump		Maximum flexibility Suitable for large systems with valves to serve multiple channels	Most expensive of the options listed Requires skilled maintenance
Tipping bucket	300–500 gal	Simple in concept May be manually operated	Rotating mechanism requires maintenance May be noise

the pit on a sufficiently frequent basis to prevent accumulation of anaerobic gases. These gases present a health and safety hazard to the livestock as well as to the workers in the building.

A variety of wastewater handling options are in use for swine systems based on flushing gutters. The most common alternatives are listed in Figures 5.7–5.11. Each of these systems incorporates one or more physical or biological treatment processes. Because these treatment processes are used with a variety of waste sources, their design is described in detail in subsequent chapters.

Flushing Systems for Dairy Cattle

Dairy cattle are maintained in various kinds of facilities, depending upon the climate of the region, herd size, the economic resources of the

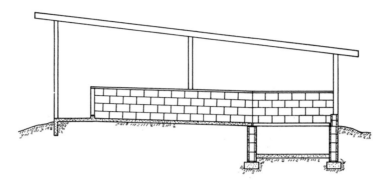

PARTIAL SLAT UNDER ROOF

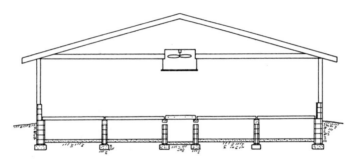

TOTAL SLAT UNDER ROOF

Figure 5.6. Two examples of the use of slatted floors to facilitate separation of animals from their manure. Manure may either be stored in the pit beneath the slatted area or may be flushed to an outside treatment or storage facility.

owner, and labor availability. A manure management system is developed in response to the type of housing adopted. One of the popular housing options, particularly for larger dairy herds, is the free-stall barn. The system typically consists of a free-stall barn in which the cows have free choice of which stall to use; a feeding area that may be separate or part of the free-stall barn; a milking parlor with a cow holding area; and frequently, a pasture or exercise area where the animals can be held for a few hours per day during appropriate weather conditions. In some systems, the cows only have access to the free-stall barn and the milking parlor for much of the year. The free-stall barn is often designed such that

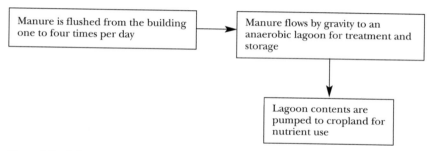

| Manure is flushed from the building one to four times per day | → | Manure flows by gravity to an anaerobic lagoon for treatment and storage |

Lagoon contents are pumped to cropland for nutrient use

Figure 5.7.   Schematic flow diagram for a simple swine wastewater handling system in which manure is flushed from the building to an anaerobic lagoon for storage and treatment prior to land application.

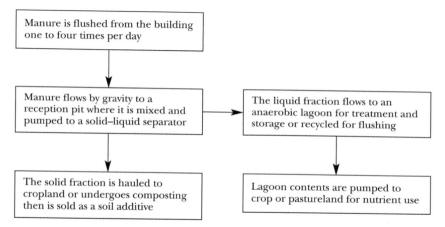

Manure is flushed from the building one to four times per day

Manure flows by gravity to a reception pit where it is mixed and pumped to a solid–liquid separator

The liquid fraction flows to an anaerobic lagoon for treatment and storage or recycled for flushing

The solid fraction is hauled to cropland or undergoes composting then is sold as a soil additive

Lagoon contents are pumped to crop or pastureland for nutrient use

Figure 5.8.   Schematic flow diagram for a swine wastewater handling system in which manure is flushed from the building with solids separated for land application or composting and the liquid fraction discharged to an anaerobic lagoon for storage and treatment prior to land application.

as the cows stand to defecate and urinate, most of the manure falls in an alleyway that is separate from the resting area by a curb 6 to 8 inches in height. If the free stalls are provided with loose bedding, a portion of the bedding also falls into the alleyway. A typical free-stall dairy barn is shown schematically in Figure 5.12. The alleyway of a free-stall barn may be scraped with a tractor blade or bucket, it may be scraped with a tractor-mounted scraper, or it may be flushed with fresh or recycled water.

Free-stall manure collection alleyways are typically 8 to 14 feet in width and may be up to 100 feet in length. Typically, if they are flushed, they have a lengthwise slope of 1 to 2% and are flat in the crosswise direction.

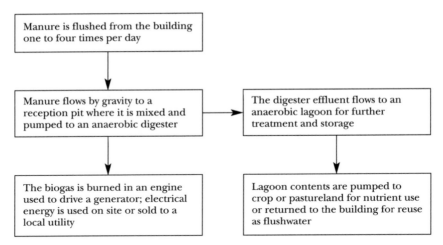

Figure 5.9.   Schematic flow diagram for a swine wastewater handling system in which manure is flushed from the building and treated in an anaerobic digester for nutrient recovery. The anaerobic digester effluent is discharged to an anaerobic lagoon for storage and further treatment prior to land application or reuse as flushwater.

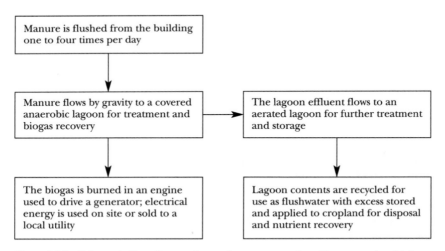

Figure 5.10.   Schematic flow diagram for a swine wastewater handling system in which manure is flushed from the building and discharged into a covered anaerobic lagoon for treatment and biogas recovery. The digester effluent is then treated in an aerated lagoon before being reused as flushwater. Excess water is applied to cropland for nutrient recovery.

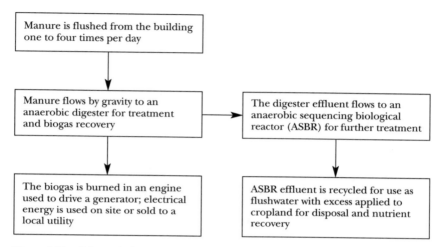

Figure 5.11. Schematic flow diagram for a swine wastewater handling system in which manure is flushed from the building and discharged into an anaerobic digester for treatment and biogas recovery. The digester effluent is then treated in an anaerobic sequencing biological reactor before being reused as flushwater. Excess water is applied to cropland for nutrient recovery.

Flushing an alley of this size takes a large water volume, hence, drop side tanks, direct high-volume pumping, or tipping flush tanks are the most common methods of initiating the flush. If the barn housing the cows is being flushed, it is also the usual practice to flush the holding area where the various herds are held in preparation for milking. Cows are usually crowded in this area and although they are held a relatively short time, considerable manure is collected in the area. Frequently, the water used to wash the milking equipment and clean the milking parlor is collected and used as part of the water to clean the holding area.

As indicated in Chapter 4, wastewater from a flushed dairy operation is relatively dilute but is produced in large volumes. Three of the typical handling methods are shown in Figures 5.13–5.15. It should be noted that liquid dairy waste handling systems are more likely to include a solid–liquid separation step than other animal manures. This is in part related to the more fibrous nature of dairy manure and the associated handling difficulty if solids are allowed to remain in the liquid.

Flushed Systems for Poultry

Flushed manure collection systems are less typical for poultry than either swine or dairy cattle. The usual flushing system for poultry is exemplified by a layer or broiler operation in which the birds are confined to cages

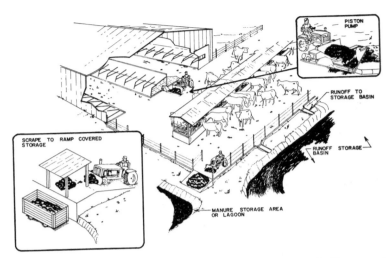

Figure 5.12. Schematic drawing of a free-stall dairy barn designed to facilitate scraping of manure from alleyways.

and these cages are mounted above a gutter that can be flushed. Typical cage design allows approximately 0.5 square feet of cage floor space per bird. The gutters are designed much the same as those described for swine. Typically, the flushed poultry droppings flow to an anaerobic lagoon for storage and treatment.

Environmental Quality Implications of Flushed Systems

The attractiveness of flushed systems is the convenience of getting the manure removed from a confinement building with a minimum of labor, having a manure that is sufficiently dilute to be handled with hydraulic equipment, and having the ability to remove manure from the building sufficiently frequently that odors can be held to a lower level than if manure is stored in the building. The disadvantages are that there is a larger volume of wastewater that requires management. This larger volume of more dilute wastewater is 10 to 50 times the organic concentration of domestic sewage, so it is a wastewater requiring careful management.

Even though dilute, many types of flushed manure cannot be handled by conventional centrifugal water pumps nor can they be controlled by valves. This is particularly true of dairy waste that is likely to contain fibrous bedding or wasted feed (grass or corn silage) that is prone to produce larger solids that clog pump impellers and collect in inappropriate places. Poultry wastes when flushed tend to contain enough feathers to create problems if provisions are not made for their removal. The

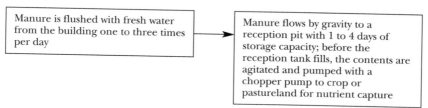

Figure 5.13.   Schematic flow diagram for the simplest of liquid dairy wastewater handling systems in which manure is flushed from the barn, collected in a reception pit, and pumped to crop or pastureland within 4 days. This system is based on having suitable land available for manure application throughout the year.

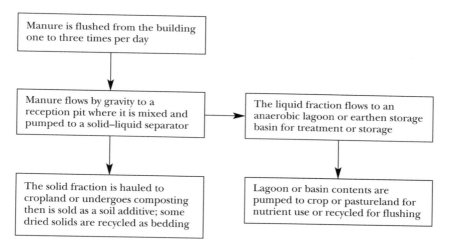

Figure 5.14.   Schematic flow diagram for a dairy wastewater handling system in which manure is flushed from the barn. The solids are extracted from the flow for recycling to land, compost production, or reuse as bedding. The liquid fraction is discharged to an anaerobic lagoon or earthen storage basin for storage and treatment prior to land application.

presence of these difficult-to-handle solids is part of the reason for the widespread incorporation of solid–liquid separating equipment in waste handling systems.

Odors are a major concern when livestock and poultry are confined and although the flushing systems have the ability to remove manure from the building more frequently than many of the alternatives, flushing alone does not ensure an odor-free operation. Flushing with recycled water can contribute another odor source to the facility, particularly if the recycled water is taken directly from an anaerobic lagoon.

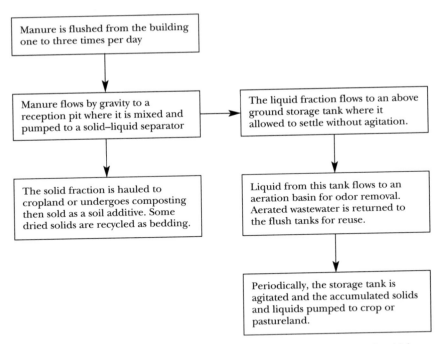

Figure 5.15.   Schematic flow diagram for a dairy wastewater handling system in which manure is flushed from the barn. The solids are separated from the flow for recycling to land, compost production, or reuse as bedding. The liquid fraction is stored in an above-ground tank. Settled wastewater from the storage tank is aerated for reuse as flushwater. Periodically, the storage tank is agitated and the accumulated solids and liquids are pumped to crop or pastureland for nutrient recovery.

In the United States, there have been very few systems designed to treat flushed animal manure for discharge to rivers or other public waters. In Taiwan, Korea, the Philippines, and other Pacific Rim countries, farms are typically smaller than in the United States and there is little land available to the pig producer for manure use so there has been a need for discharging systems. As designers of waste treatment systems, this is a particular challenge. Wastewater treatment becomes increasingly expensive as the percentage of pollutant removal approaches 100. Anaerobic lagoons or enclosed anaerobic digesters are discussed subsequently in this book, and it will be clear that achieving a 90% BOD removal with these processes is possible. Achieving the last 10% costs at least as much as the first 90% and may require an even greater investment. Even with these good removals, effluent concentrations are still high because animal wastewater is so concentrated. Similar logic is also applicable to odor control. It is possible to decrease the intensity of odors from a livestock confinement facility by appropriate design and management. Total elimination of the odors is considerably more expensive.

Manure Slurry Systems

Manure as excreted by most livestock is between 80 and 90% water. At this moisture content, it is not easily handled with either solids handling equipment, pitchforks, or front-end loaders; or with liquid handling equipment such as pumps, gate valves, and hydraulic conveyances. There is an inherent advantage of handling manure at its minimum volume and weight. This requires the amount of added water be minimized. Systems are in use that are based on this concept and that are considered fully satisfactory in several situations.

Slurry Systems for Swine Production

There are multiple examples of the use of manure storage tanks with confinement swine production. These tanks may be located beneath the building, adjacent to, or at some distance, depending upon the local climate, soil conditions, and overall plan of the facility.

Underfloor storage pits. The most obvious application of this concept is the underfloor storage pit, which is widely practiced throughout the swine industry. In this system, a totally or partially slatted floor is constructed over a pit area that may be under 30 to 90% of the building. The depth of the pit varies depending on the soil conditions of the area, depth to groundwater, and desired storage duration. Typically, storage pits are designed to provide 5 to 12 months of storage. Although the design of storage pits is a relatively straightforward matter of multiplying the volume of waste material produced per pig each day by the number of animals and the days of anticipated storage, accurately predicting the volume of waste material per pig is difficult. The waste material not only includes the urine and feces but also any wasted feed and water, as well as any water used to clean the pens or cool the pigs. In addition to the storage volume of the pit, an additional depth of at least 1 foot should be allowed as freeboard between the maximum manure height and the bottom of the slats. This freeboard should be increased at least an additional 6 inches if plans include provision of a pit ventilation system.

Pit ventilation is drawing the minimum ventilation rate for the building from a series of inlets located in the manure storage pit, beneath the slats, and above the maximum manure storage level (Figure 5.16). Pit ventilation removes the carbon dioxide, methane, hydrogen sulfide, and

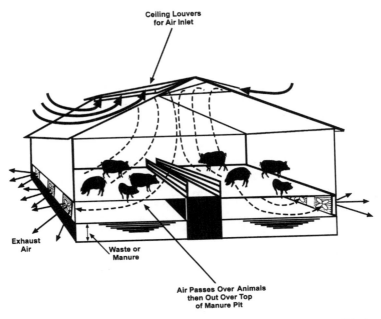

Figure 5.16. A pit ventilation system in which air from the manure storage pit is drawn from above the manure surface but beneath the slats to minimize to presence of manure-produced gases in the animal housing area.

ammonia released by the decomposition of the manure before it has an opportunity to rise into the living space of the pigs. Pit ventilation systems are not 100% effective in keeping the released gases out of the building but most operators seem to think they are beneficial.

Confinement buildings with underfloor manure storage pits are critically dependent upon the continuous functioning of the ventilation system. It is appropriate to include in the design some kind of warning system so the operator can be alerted at any time whether there is a power failure or other event causing the minimum ventilation fans to cease operation. Animals can be killed within a few hours. When a power failure occurs, all doors and windows should be opened or alternatively, the generator associated with the back-up power supply should be activated.

In addition to the ventilation aspect of an underfloor manure storage pit, it is also important to consider how manure is to be removed from the pit. Stored manure requires agitation before it can be pumped from a pit. Therefore, in designing the pit it is important to know the way in which agitation will be accomplished. Agitation is necessary to homogenize the manure so each load is of the same nutrient composition and to remove all of the manure from the pit. Experience with removing manure without agitation is that a relatively small cone of air develops around the pump intake and primarily liquid is removed, leaving most of the manure solids behind and effectively reducing the volume available for future storage. One option is a manure chopper pump that has a discharge beneath the manure surface as well as to the vehicle being used to haul the manure from the building to the land application site. It is also possible to purchase a manure transport vehicle that is equipped with a pump that creates a vacuum in the tank that can pull manure slurry from the storage pit into the tank. That same pump also can be used to create pressure in the tank, forcing the first liquid removed back into the pit, creating a measure of agitation. Whatever the mode of agitation, the pit needs to be designed to allow agitation of all of the manure to remove it. As a result, multiple access ports are necessary along the length of the pit. The space between these ports is selected based on the effective reach of the agitation equipment.

Exterior manure storage tanks. Swine manure can be stored in tanks outside the building just as well as in tanks beneath the building, however, the former storage is far less common. One option is the deep, narrow pit in which a pit is constructed beneath a slotted section of a partially slotted floor pen design. Common practice is to install a drain outlet in the bottom of such a pit and to fit an upright pipe into the drain opening to control the water level at the top of the standpipe. In this system, the standpipe is periodically pulled from the drain and the manure slurry drains by gravity to an outside storage tank. This design yields manure somewhat more fluid than fresh manure but with considerably less volume than produced with a flushing system. Odor within the building is one of the disadvantages of this system.

One option for storing manure from a deep, narrow pit system is in an earthen storage basin. Earthen storage basins are generally constructed in the soil by excavation and compaction so they are partially below grade and partially above. They need to be constructed to avoid both surface and groundwater pollution. By having the dikes constructed of the soil excavated for the basin, the inflow of surface water is generally avoided. It is desirable to construct the basin such that the design high water level is just below the drain elevation of any pits that are to be served so gravity flow can be possible. Most of these basins are designed to serve smaller swine production facilities and assistance can be obtained from local Farm Service Agency personnel in determining site suitability, dike construction, and interior sealing. It is important that these earthen basins be watertight to prevent groundwater pollution. It is common to construct a concrete access ramp from the top of the dike down to near the basin bottom to allow easy tractor access for emptying the basin with a power take-off driven pump. Table 5.18 lists some of the design features that are important. Figure 5.12 shows a typical manure storage basin as might be used for a free-stall dairy barn.

Safety Note!

It is important for both the designer and the operator to understand that the gases released by the decomposition of manure can kill animals and operators. Large quantities of these gases are released when the manure storage pit is agitated. Thus, it is critically important that either the animals be removed from the building when the manure is being agitated or that the agitation be scheduled at a time the building can be fully open and ventilated.

Hydrogen sulfide is a toxic gas. It kills both livestock and people. Carbon dioxide, methane, and ammonia all replace oxygen and can cause asphyxiation. As a result, it is highly dangerous for people to enter manure storage tanks. The concentration of gases is likely to be sufficiently great as to cause a person to immediately become unconscious, collapse, and drown. In the design of all manure storage tanks, operator safety is a prime concern.

Protective guardrails and a cover over access ports to protect even small children should be in place at all times. This is a point of demanding design. The covers should be designed so they prevent accidental entry, as well as the entry of children or pets, but are sufficiently convenient that operators do not permanently remove them because they are a hindrance to normal operation.

If it is necessary for anyone to enter a manure storage tank, rigorous safety precautions are necessary. One option is to have the person fitted with and experienced in using self-contained breathing equipment such as that used by deep-water divers. Other options are to ventilate the tank extensively before the person enters; have the person fitted with a rescue harness; and have a person or persons available at the entrance to remove the person immediately upon any sign of weakness, fainting, or light-headedness. Above all, should a person in a tank collapse, the person at the tank entrance must avoid the temptation to enter and attempt a heroic rescue. The literature is already oversupplied with examples of multiple deaths associated with manure storage tanks.

Table 5.18.   Design details for the construction of an earthen manure storage basin

Place the manure storage basin where it is not likely to be an eyesore or create odor complaints.

Evaluate the soil characteristics to be certain that an impervious basin can be constructed. Clay materials work best.

Evaluate the site to ensure that the basin bottom can be placed at least 3 feet above bedrock and at least 2 feet above the maximum high-water table.

Construct the dike at slopes that can be maintained. The outer dikes should be sloped at 3 or more horizontal to 1 vertical to facilitate mowing. Steeper interior slopes up to 2 to 1 are desirable if the soils can maintain this slope.

The dike should have a top width of at least 12 feet to facilitate driving along it and for mowing.

Based on site and soil conditions, it is desirable to have the basin as deep as possible. This minimizes the surface area-to-volume ratio and reduces odor and rainfall input. This greater depth also facilitates the formation of a dry crust, which is desirable for odor control.

If a concrete access ramp is to be constructed, design the slope to accommodate the transport vehicles you anticipate being used. If a tankwagon is to be used, restrict the slope to 10%. If only a tractor and pump access is required, a slope of up to 5 to 1 can be used. Concrete ramps should be formed with horizontal grooves to provide traction even when wet and with a manure-spattered surface.

Two other possibilities should be briefly mentioned herein. One is the use of an aboveground manure storage tank; the other is anaerobic digestion in a sealed tank to preserve nutrients, capture biogas, and partially treat manure for more convenient application to crop or pastureland. Aboveground storage tanks can be constructed of a variety of materials and are managed much like belowground tanks except that they can be larger and can be designed to facilitate a more sophisticated manure treatment system. Because aboveground manure storages are more typically used on dairy farms, they are discussed in full detail along with the other dairy manure options.

Manure slurry is of an appropriate concentration for anaerobic digestion. Manure of 5 to 10% solids is thick enough to make efficient use of digester volume, yet sufficiently dilute that it can be handled with appropriately sized and selected manure handling equipment. An anaerobic digester is shown conceptually in Figure 5.17. A typical flow pattern for a manure handling system, including an anaerobic digester, is outlined in Fig. 5.18.

Slurry Systems for Dairy Cattle

There are several alternative manure handling systems in use to collect dairy cattle manure that result in a slurry with a solids content between 5 and 12%. The simplest is the scraped free-stall system in which the animals drop the bulk of their manure in an access alley between rows of free stalls. While the cows are in the milking parlor or holding area, a small tractor or front-end loader fitted with a blade is used to scrape the

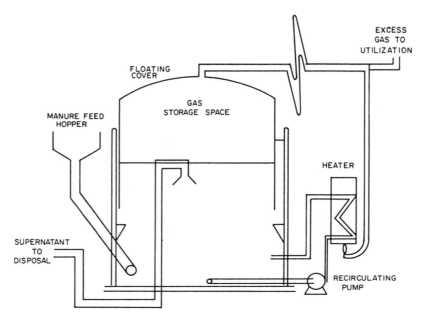

Figure 5.17.   Conceptual design of a floating cover anaerobic digester.

manure into a grating covered pit. Washwater from the milking parlor as
well as manure scraped from the parlor and holding area also are placed
in this receiving pit. As the pit approaches capacity, an agitating pump is
activated to homogenize the collected manure and washwater. After being homogenized, the manure slurry is pumped into an aboveground
storage tank. Seasonally, the aboveground tank is agitated and manure
hauled to cropland, typically with a tank wagon or truck-mounted tank.
This system handles manure in a minimal volume, conserves a large fraction of the nutrients, and does not require an irrigation system. Routinely, the labor requirements of this system are not overly demanding but
if the herd is very large, manure hauling to fields is a time-demanding
operation.
    An alternative to this system is to use an underground storage tank or
a manure storage basin as described for slurries from pig production
(Table 5.13). These systems are particularly popular for dairy herds of
100 or fewer cows. An outside, uncovered manure storage basin is shown
in Figure 5.19. In these systems, manure can be scraped from the barn
with a small tractor or may be moved to storage with a scraper system incorporated in the floor, which moves slowly through the building in a

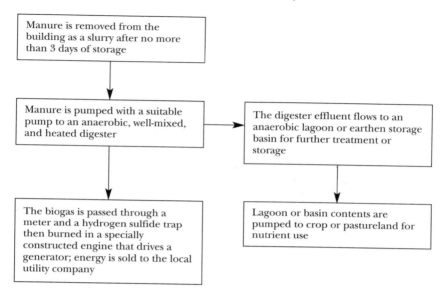

Figure 5.18.   Schematic flow diagram for a swine or dairy slurry waste handling system in which manure flows from the barn and is pumped into a heated, well mixed anaerobic digester for biogas recovery. The digester effluent is applied to cropland for nutrient recovery.

shallow gutter. Movement is sufficiently slow that it does not interfere with cow traffic in this area. The scraped gutter is behind the resting stalls so it does not interfere with resting cows.

A third option is a free-stall barn for a larger herd in which the alley behind the cows is a slotted floor area. Animal traffic forces the manure through the floor openings into a shallow gutter, similar to one that might be flushed but usually at less of a slope. Periodically, usually twice daily, a cable-pulled scraping mechanism drags and pushes accumulated manure the full length of the gutter into a deeper but more narrow cross-collection gutter where it is either scraped or flushed to a common collection point. From this collection point, the manure can flow by gravity into an earthen storage basin, an underground storage tank, be pumped into an aboveground storage tank, or can be subjected to anaerobic digestion as described in Figure 5.18. Anaerobic digester systems have proven more popular for dairy cows than other animals because of the on-site need for hot water, which is a necessary by-product of producing electricity by this process. The anaerobic digestion process is described in greater detail in a subsequent chapter.

Figure 5.19.   An outside, uncovered manure storage basin as might be used with a manure scraping system.

Slurry Systems for Beef Cattle

Beef cattle are typically finished in cattle feedlots as described previously in this chapter, but there are several operations around the country in which beef cattle are being housed in long farrow barns with a slotted floor section running the length of the building. These buildings are frequently 400 feet in length and 30 to 40 feet in width. They are often built in pairs with a cattle-working lane in between the two barns. In some of these systems, the manure is removed from the gutter with a scraper mechanism similar to that described for dairies. The slurry of manure scraped from under the slats can be stored as a semisolid in an earthen basin or in more arid areas, it can be dried in the sun and either composted along with other materials or spread as a solid manure.

Solid Manure Handling Systems

Traditionally, manure was handled by the addition of sufficient bedding to an animal housing area to absorb the urine and extract enough of the moisture from feces to allow a relatively low-odor and healthy environment for the animals and a pleasant work area for the farmer. These systems still exist under modified conditions. For example, recreational horses are frequently bedded in stalls into which straw or some other absorbent material is placed to dry the manure and urine. This system maintains a clean animal and prevents many of the health problems associated with fecal contamination. In many areas, the horse manure, straw, feces, and urine combination is in demand both as a home gardening soil amendment and as a starting material for the compost used by commercial mushroom growers.

Solid manure handling has several advantages for avoiding water pollution, however, it tends to be associated with labor-intensive systems. The exception is in the poultry industry in which highly labor-efficient systems have developed that yield a solid material. From a water quality perspective, these systems provide the potential for a high degree of protection. The challenge to most of the solid manure-producing systems is appropriate storage of the manure after it is removed from the animal facility and its safe use either as a soil amendment or for one of the other uses that have evolved.

Solid Manure Handling in the Poultry Industry

There are three important poultry production systems that produce a dry manure. All three are widely used and can lead to a low risk of environmental damage. All three, however, require both careful design and a high degree of operator attention to managing the system.

Deep pit systems. Deep pit systems, sometimes called high-rise buildings, are used extensively for egg production and for raising pullets. The birds are typically maintained in cages at densities of two to five birds per square foot. The cages may be single layer or arranged in a stair-step fashion. The cage area is elevated so there is functionally a lower level beneath the bird-housing area. The droppings fall into this lower area. The height of the lower area is dependent on the intended method of removing the manure. If a tractor with a front-end loader bucket is intended, sufficient space must be allowed for it to maneuver. Figure 5.20 shows a deep pit system for the storage of poultry droppings.

The concept involved in a deep pit building is that the droppings are dried and stored in a sufficiently dry condition to minimize odor production. For this system to operate successfully, there must be adequate air movement through the building to remove the latent heat produced by the birds and to maintain a manure layer that has a moisture content of less than 50% on a wet weight basis. If wet spots develop or if the entire accumulation becomes wet, the area becomes a severe odor problem and the ammonia release poses a major threat to both the birds and the neighbors. For this not to happen, three components must be in place.

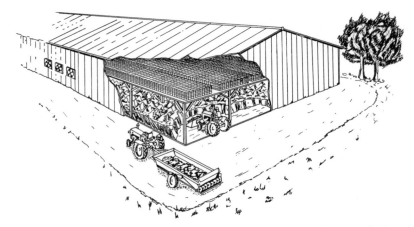

Figure 5.20.   A deep pit system for the storage of poultry droppings. This system is also called a high-rise poultry system because the building is usually taller than typical to accommodate both solid storage and the bird housing area.

1. All extraneous water must be kept out of the manure storage pit. Thus, the watering system, including the drinking cups, must not leak. In most buildings, this means not only careful selection during construction but also an ongoing inspection and repair program to respond immediately to overflowing cups, dripping nipples, or leaking valves and pipe joints. It is also appropriate in many areas to construct a tile drain around the perimeter of the building that drains to an outlet or to a pumped sump to ensure that high groundwater does not enter the storage area.

2. Exhaust fans are installed along the length of the building with sufficient capacity to exhaust the evaporated water. Air inlets are to the upper portion of the barn. Thus, any ammonia or odor released by the droppings is exhausted from the storage area rather than being allowed to pass through the bird-housing portion of the building. Supplemental heat is required under many conditions to maintain appropriate temperatures for the birds while simultaneously providing exhaust fan operation to maintain the manure sufficiently dry. A carefully done heat and moisture balance identifies the amount of supplemental heat necessary. Experience indicates it is important that sufficient supplemental heat be available because if the pack becomes wet, removing the excess moisture is a long and frequently difficult process.

3. Circulation fans placed in the lower portion of the building to circulate air over the droppings is essential as a way to fully use the ventila-

tion air and supplemental heat being provided. Circulation fans are commonly installed approximately every 100 feet to maintain a continuous air movement in an oval pattern throughout the storage area.

Frequent monitoring of the manure storage area is necessary if odors and wet manure are to be avoided. Physical inspection and sampling tell the operator if leaks are occurring from the watering devices and if overall evaporation is sufficient. In difficult buildings or in buildings located in proximity to residences, it may be necessary to put down a layer of dry absorbent material before birds are put in the cages to promote drying of that first layer. When wet spots are discovered, it is desirable to remove all the material from that area. Ammonia-absorbing materials such as natural zeolite also may be spread on moist areas to help retain the ammonia and to lower the moisture content.

If a deep pit system is operating satisfactorily, manure accumulation is slow and the odor level within reason. Some operators talk of being able to wait for up to 3 years to empty the storage areas, however, annual manure removal is more typical. If moisture enters the building due to high groundwater or widespread leakage, immediate manure removal is likely to be necessary. Dry poultry manure is a quality fertilizer and after analysis can be applied to cropland, it can be composted with other more carbonaceous material to produce a high quality soil amendment, or it may be sold directly. Processing and bagging of poultry manure for the consumer market is widespread. An important feature of systems for which these operations have been successful has been marketing. Anyone contemplating purchasing grinding, mixing, and bagging equipment would be well advised to conduct a careful market analysis prior to investing. Only a few of the many poultry manure marketing ventures have proven financially successful.

Cage systems with dropping boards. An alternative to the deep pit system is a poultry housing system in which the cages are layered with a dropping board between the layers. In this system, the droppings fall directly onto the dropping board for either partial or total drying. Periodically, every 1 to 3 days, the droppings are scraped off the boards either to a shallow pit, the floor, or into a small cart, and hauled to a covered storage area for further drying or stabilization. This material is a high-quality soil amendment or it can be used as a nitrogen source in an integrated compost production operation. Extensive experiments also have been conducted documenting the value of this material, after further processing, as a feed ingredient for ruminants or even for incorporation in poultry rations. For a variety of reasons, some economic and others aesthetic, this practice has not been widely adopted in the United States, although range coprophagy has long been recognized as a significant nutrient source of chickens following cattle. Refeeding options are further discussed in Chapter 11.

Litter on floor operations. Most broilers, breeding birds, and brooding operations are conducted on litter-covered floors. The buildings are typically single story with a dirt floor. In the mild-climate regions, they are likely to be open-sided with drop curtains to conserve heat but designed to be opened when it is warm to achieve maximum ventilation. In the areas with cold winters, the buildings have solid walls, insulated roofs, and mechanical ventilation. Concrete floors are also common in the northern states. Self-feeders and automatic watering devices are mounted throughout the building so the birds have constant access to feed and water. Supplemental heat is provided in areas where needed. Brooder hoods either gas-fired or electrically heated are commonly provided to keep freshly hatched birds warm and dry.

The usual manure management practice is to place 6 inches or more of dry, loose litter on the floor before birds are first admitted and then to not disturb the litter through the growing cycle. Between flocks, the litter may be stirred with a rototiller-type tool and additional litter may be added if necessary. Mixing and stirring the litter promotes a natural composting process. One challenge facing the manager of this type of system is to manage the litter so it is sufficiently moist to prevent excessive dust and also sufficiently dry to avoid anaerobic biological activity that can lead to release of ammonia at a level harmful to the birds. Proper selection, monitoring, and repair of watering devices are critical. Watering cups, nipples, and open troughs are all in use and can be managed to avoid spills. There are strong advocates for each of the watering devices.

Periodically, often once a year but more often if necessary, litter is removed from the building. This is usually done with a front-end loader. The removed litter can then be stockpiled, hauled directly to crop or pastureland, or composted to improve its handling characteristics. The amount and quality of litter removed from poultry buildings is highly variable depending on the amount of litter used, how long it is allowed to remain in the building, and the extent to which droppings were composted while on the building floor. In general, however, a large portion of the dry matter and the nitrogen is either volatilized or exhausted as dust. Thus, it is interesting to compare the characteristics of manure removed from a deep pit with the used litter from a floor-rearing operation. Table 5.19 provides a summary of typical values. The message in this table is that both materials are likely to be highly variable in composition and if precision management is to be achieved, on-site analyses are necessary.

Table 5.19. Typical values, amount and composition of deep pit poultry manure and used litter from solid floor rearing facilities

Characteristic	Value for manure removed from a deep pit poultry facility	Value for used litter from a floor-rearing poultry operation
Mass, lb/1,000 lb per day	23	28
% total solids, (TS)	55	72
TS, lb/1,000 lb per day	11	20
Total nitrogen, % TS	3.5	3.5

Solid Manure Handling Systems, Other Than Poultry

There is a lot of solid manure generated in the United States and throughout the world. Many of the systems generating solid manure are small, thus the labor involved is not a serious concern. Horses in the United States are largely maintained as recreational animals and the stalls are freely bedded with straw, wood shavings, or other absorbent materials to keep the animals clean and dry. Management consists of removing the soiled bedding on a regular basis and replacing it with clean material. If it is to be stored before disposal, a roofed storage is desirable to prevent runoff from extracting the soluble nutrients, organic matter, and indicator microorganisms, and carrying them to nearby surface waters. Extreme nitrate loads also can develop below an unroofed manure pile. If a roof is impractical, covering with a waterproof fabric cover provides reasonable protection to both surface and groundwater.

Larger solid manure-generating facilities generally involve some mechanical means of removing the manure from the barn. This may be a mechanical barn cleaner, a tractor-mounted manure loader or scraper, or scraper and piston pump to move the manure from a collection point to a storage area. Two concerns of the designer are the convenience the system affords the operator and environmental protection. For the operator, minimizing labor is important. If manure is to be moved with a tractor-mounted bucket, a bucking wall is helpful. The storage area should be conveniently located and provided with all-weather access. Environmentally, the goals are to prevent runoff and infiltration. Surface water should be drained around rather than through the manure storage, and any liquid from the manure storage should be collected or otherwise precluded from stream discharge. A roofed storage area is necessary in many regions. In designing the roof, it is important to allow for the full height of any mechanical equipment to be used in the handling of the manure.

Deep Litter Manure Handling System

An alternative to liquid systems that has been used in various locations around the world is the deep litter system. In this system, the animals are confined on a floor covered with 2 to 3 feet of sawdust or shavings mixed with manure. Both swine and poultry have been raised on this system. The management is to periodically, once or twice weekly, mix the freshly deposited manure and urine with the bedding material. To function properly, the ventilation rate must be sufficient to remove the moisture from the area so the moisture content of the bedding material–manure mixture remains between 40 and 60%, which allows air to permeate the system and support a composting operation. This system has a higher labor-input requirement than some of the other alternatives, but it has the potential to prevent water pollution, produce a useful stable byproduct for land application, and provide a clean, healthful environment for the animals.

References

Midwest Plan Service. 1985. Livestock Waste Facilities Handbook, MWPS-18. Ames: Iowa State University, MWPS.

Miner, J.R., R.B. Wensink, and R.M. McDowell. 1979. Design and Cost of Feedlot Runoff Control Facilities. EPA-600/2-79-070. Ada, OK: U.S. Environmental Protection Agency.

Voermans, J.A.M. and C.N. Huysman. 1990. Decomposition of pig manure by using additives in combination with deep litter of sawdust and shavings, pp. 20–24, In: Proceedings of the Sixth International Symposium on Agricultural and Food Processing Wastes. ASAE Publ. 05–90.

6

Solid–Liquid Separation

One of the first problems encountered by the designer of livestock and poultry waste handling facilities is that the material to be handled does not behave the same way as water, solid grain particles, or municipal wastewater. Animal manure as excreted is a mixture of urine and fecal material. By the time it reaches the manure handling equipment, it may be augmented by wasted feed, soil particles, surrounding vegetation, animal hair or poultry feathers, extraneous biological materials such as afterbirth and stillborn fetuses, soil particles, and additional water. This additional water may arise from the drinking water system, from the automatic flushing or the manual cleaning of pen structures, or from rainfall as in feedlot runoff. Animal fecal material is also variable depending upon the animal species and the physical form of the feed. Obviously, the feces of dairy cattle depends upon whether they are fed a coarsely chopped corn silage or a pellet prepared from finely ground concentrates. All of these variables are important to the handling characteristics of waste material. Frequently, the designer decides that the overall system may be more effective, less expensive, or otherwise more acceptable if solid–liquid separation is incorporated into the system. Table 6.1 lists some of the more typical situations that lead to the incorporation of separation processes.

There are a variety of reasons for considering the incorporation of solid–liquid separation equipment into an overall livestock waste management plan (Table 6.1). These reasons can generally be summarized as follows:

1. To reduce the maintenance requirement of subsequent storage and treatment operations.
   a. Removing larger and particularly fibrous materials from a liquid waste stream reduces plugging problems, particularly for systems that involve small-bore irrigation nozzles, precision pump impellers, or mechanically controlled flushing equipment. Chicken feathers are notoriously effective in plugging valves or nozzles.
   b. Runoff from feedlots is captured in runoff retention basins. Runoff also carries large quantities of solid material, such as soil and

121

Table 6.1. Examples of animal waste handling situations that suggest inclusion of some form of solid–liquid separation

Source of material	Nature of the solids	Problem to be avoided by solid–liquid separation
Runoff due to rainfall on an unsurfaced feedlot.	Soil particles are carried by the runoff as well as manure solids consisting of undigested feed, hair that was shed by the animals, and trash vegetation that may have blown onto the feedlot.	Soil particles in particular are inert and if allowed to enter the runoff retention basin, remain on the bottom occupying space that would otherwise be recoverable for runoff storage. Subsequent dredging of a runoff retention basin is a relatively expensive process in many locations.
Runoff due to snow melting on a feedlot surface.	Snow melt runoff tends to be relatively low in volume but may contain up to 10% solids. The solids are similar to those in rainfall caused runoff.	Large volumes of solids and soil particles consume the storage capacity of the runoff retention basin and can more inexpensively be removed in other facilities.
Flushed wastes from a dairy barn.	Depending upon the nature of the ration, solid material may range from discrete grain particles to long strands of fibrous grasses or corn silage. Dairy manure solids have value for reuse as a bedding material or as a potting mix if appropriately processed.	The waste as flushed from the building is likely to be difficult to pump through conventional irrigation equipment without plugging the nozzles or becoming entangled in the impeller. If the flushwater is to be placed in a storage tank, the solids are likely to settle to the bottom or float to the surface and be difficult to resuspend for later removal.
Liquid waste from a poultry operation, either flushed from a channel under cages or dumped from a pit within the building.	Liquid poultry wastes contain feathers, and broken eggs in addition to droppings.	Feathers are relatively slow to degrade and provide an abundance of pump impeller and valve plugging problems.
Dilute swine waste from a flush system.	Depending upon the physical nature of the feed ration, the manure particles may be from discrete grain size to much smaller particles.	Solids separation reduces the organic and nutrient load on downstream biological processing systems. Solids removal also removes extraneous solids that may have entered the manure system.
Treatment process, such as digester or lagoon effluent, that is destined for recycling or discharge.	Floating or suspended biological mass such as clumps, bacterial cells, or foam.	Biosolids left in an effluent may cause downstream pollution problems, may contribute to odors, or may have value if reclaimed and concentrated.
Lagoons or storage basins that become filled with solids.	Solids in lagoons that become filled are a mixture of manure solids and bacterial cells. Runoff collection basins also may contain soil particles filling space needed for treatment.	Removing solids restores the unit to appropriate functioning condition. Trailer-mounted centrifuges have been used for this purpose as have sand filters and sand drying beds.

manure particles. If particulate material enters the runoff retention basin, it can occupy space intended for liquid storage. This solid material is often expensive to remove, so designers have learned it is more effective to capture the solids before they enter the basin.

2. To reduce the organic and nutrient load on subsequent biological treatment processes.

   a. Lagoons as well as other biologically based treatment devices, including aeration tanks and constructed wetlands, are designed in part based on the organic load to be treated. It may be cost-effective to remove a portion of the load with a solids separator. An alternate but similar logic is that by separating the solids prior to placing liquid manure in a lagoon, the load is reduced. Thus, the lagoon can operate at a lower loading rate, resulting in reduced odor, or the size can be reduced, also resulting in a decreased odor-releasing surface.

   b. There are also livestock and poultry operations where the land available for nutrient recovery limits the operation's size. Solids removed from the incoming waste stream may reduce the effluent nutrient concentrations sufficiently to allow expansion or to more adequately accommodate the existing operation.

3. To claim the solid fraction for a beneficial use that would otherwise be missed.

   a. Most dairy rations include a high-fiber feed such as hay, haylage, or silage. If this material is separated from the fresh waste stream, dried, and composted, it makes an acceptable bedding material. With the cost of bedding material increasing in many areas, this practice has become more attractive. Incorporation of separated solids into the potting mix of plant nurseries is another potential beneficial use. The separated solids provide a slowly decomposing organic material with a relatively large water holding capacity.

   b. Some swine system operators use solid separation equipment to recover grain particles from finishing building flushed wastes and recycle them as a component of the ration for gestating sows. This material not only reduces the feed cost for energy and protein but also appears to reduce disease through the immunization effect of recycling the material back to the herd.

4. To reclaim the useable function of a tank, basin, or lagoon that has become filled with solids.

   a. Frequently, in spite of reasonable design efforts, a tank, basin, or lagoon becomes filled with solids and no longer functions as planned. In such cases, it must be restored. One alternative for restoration is to pump the contents out and through a solid–liquid separator. The liquid can be returned to the unit and the solids applied to crop or pastureland.

b. Undersized anaerobic lagoons are particularly prone to becoming filled with solids. One option to restore them is to pump the contents through a truck-mounted centrifuge (Miner et al. 1983). This approach was applied in Singapore where pig farmers were plagued with severe land shortages, hence they could only build very small lagoons that tended to function more like settling tanks. They became solids filled after 1 to 3 years. They were restored by harvesting solids with a portable centrifuge. The lagoon solids were much more amenable to centrifugation than was fresh manure.

Solid–Liquid Separation Processes

There are two characteristics of solids in liquids that are used as the bases for separation of solids from liquids. One is based on the density of solids being different than that of the surrounding liquid. Under quiescent conditions, dense particles settle to the bottom of a container containing liquid that is less dense. From physics, we learned that the buoyant forces on a submerged body are equal to the weight of the fluid displaced. As long as the particles are more dense than the surrounding liquid, they have a net downward force. Clearly, more dense particles settle faster than those whose density is closer to the surrounding fluid and similarly, larger particles settle more rapidly than smaller particles. Dry particles with entrapped air do not settle or at best settle slowly. Hence, a manure slurry with a lot of straw and other cellulosic material containing small pockets of air does not settle but instead floats.

The other character used to separate particles from liquids is their physical size. Consider how a screen or tea strainer works. The tea leaves are at least one dimension larger than the openings in the screen. As soon as a layer of tea leaves forms on top of the screen, the leaves actually do the straining and frequently they strain out particles that are actually small enough to pass through the screen, but they do not because they never get to the screen. Similarly, a strainer that becomes covered with captured leaves soon ceases to allow liquid to pass through at an acceptable rate. Early efforts to separate manure solids from liquids attempted to use a plain screen and it did not work well because the manure articles quickly covered the screen openings and the screen was effectively plugged. Current screens overcome this problem by incorporating a feature that moves the solids off the screen soon after they are trapped.

Before considering the facilities and equipment to separate livestock and poultry waste solids from liquids, it is important to consider the nature of the solids likely to be removed. Fresh animal manure is a mixture of undigested feedstuffs, by-products of digestion, and a large bacterial population. All of these components tend to be particles, solutions, and emulsions that bond with water and drain only slowly. In contrast, if the manure is mixed with several volumes of water as is typical of flushing systems, then subjected to solid–liquid separation, the separated solids behave very differently. Most likely, they do not resemble fresh manure even though they may have the same moisture content. This phenomenon is most dramatically demonstrated by the separation of dairy manure solids from a flushing operation. Fresh dairy manure as defecated is a mixture of materials that dries slowly because water is tightly held by a mixture of different-sized particles. In contrast, if a flushed dairy manure is subjected to screening, the solid fraction is a pile of fibrous material that drains easily, does not attract flies, and tends not to have the typical manure odor. Thus, in planning a solid–liquid separation process, it is important to give some thought to the nature of the solid that may result.

Separation Based on Particle Density

There are many natural examples in which solids are separated from liquids because the solids have greater density than the supporting liquid. The one closest to our application is a stream immediately after a storm; the stream carries a large sediment load (indicated by its brown color). If this stream flows into a lake or pond, the solids settle and the pond or lake becomes clear again. Similarly, after a few days, the slow-moving water in the stream is clear again, but there is a layer of sediment on the bottom of both the stream and the lake.

There are two types of settling devices in use for the treatment of livestock and poultry waste. One is best described as a settling tank in which there is a fixed water level and solids-laden water flows in, is retained long enough for the solids to settle, and then the clarified liquid flows out because more liquid enters. This kind of settling tank works best for systems that have continuous or regular wastewater flow. The usual design basis for a tank of this type is to size the tank to have a specified residence time or even better to have a liquid-loading rate based on the flow rate divided by the bottom area of the tank. Settling tanks for domestic sewage treatment are typically designed to have a residence time of 1 to 2 hours. Settling tank experience for livestock wastes suggests a settling tank having a surface area of approximately 900 square feet per cubic foot per second flow rate. This is equivalent to a more standard expression in the waste-

Table 6.2.   Performance of a settling tank with 30-minute detention time receiving 1.2% total solids (TS) flushed swine manure

Characteristic	Performance, %
TS removal	35
BOD removal	25
Total nitrogen (N) removal	15
Ammonia N removal	10
TS in the bottom sludge	7

water field of 900 gallons per day per square foot. It is possible to use this latter number by calculating the daily flow in gallons per day and then dividing by 900 to determine the appropriate surface area. It is important in considering a tank of this type to remember that the accumulated solids must be removed on a regular basis or they begin to decompose and gas is produced that floats solids to the surface. In municipal wastewater treatment, these solids are routinely scraped to a hopper and pumped out several times per day.

Characterizing the effectiveness of settling tanks for the treatment of animal manure is difficult because such different materials are sent to the tank. The estimates provided in Table 6.2 are based on a settling tank with an average retention time of 30 minutes receiving a flushed swine manure of approximately 1.2% total solids.

The second type of gravity settling unit is one designed to respond to storm events such as feedlot runoff. In this unit, the goal is to capture solids that would otherwise flow into a larger storage unit and effectively reduce its storage capacity because of the difficulty of removing accumulated solids from a large pond that is never dry. Frequently, it is possible to incorporate a settling basin in the runoff collection system by having a channel section with low velocity but sufficiently wide enough that solids are deposited. The usual design approach is to select an appropriate design storm intensity and use this intensity as the basis for a flow rate. The 10-year, 1-hour storm intensity is frequently selected. This is a widely available weather statistic (Midwest Plan Service, *Livestock Facilities Handbook*). Knowing this intensity and the area draining to the system, a flow rate can be calculated in cubic feet or cubic meters per second. A basin large enough to provide 30 minutes of settling time adequately removes solids from runoff. Thus, with a few calculations the volume needed can be determined. A solids-capturing basin should be relatively shallow to facilitate settling and to preserve the elevation head necessary for the liquid fraction to flow into a subsequent storage basin. A depth of 2 to 3 feet is typical. Figure 6.1 shows a typical earthen settling basin.

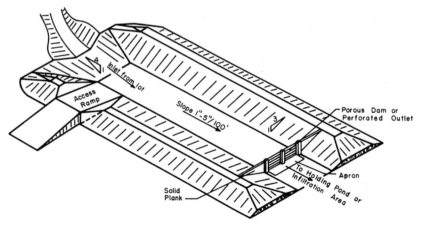

Figure 6.1. An earthen solids settling basin as might be constructed to retain solids carried with open feedlot runoff.

The details of a solids removal basin are usually dependent on the local topography but in each case, the designer needs to be concerned about how the liquid fraction may be drained away and how the solids may be removed from the basin. One method for allowing the liquid to drain out of the solids trapping basin is to use a porous dam that is designed to trap water but subsequently allows it to slowly pass. These porous dams can be built of horizontal, wooden planks with space between them (0.5 inch) for water to flow. V-notch weirs are useable as are weirs with a narrow, vertical opening in the lower part and a wide, horizontal area near the top to allow maximum flow to pass in case of a storm event of greater than the design intensity. Depending on the nature of the solids collection basin and the frequency of storm events, it may be appropriate to provide a paved access ramp into the basin to facilitate solids removal. Retention basins also frequently have paved bottoms to facilitate equipment working in them to scrape and load the accumulated solids.

Design data on the performance of settling basins to reduce the solids entering a runoff retention basin are more variable than those for settling tanks after flushed systems because of the variability of the influent. Typically, a settling basin would be expected to trap and retain a higher fraction of the total solids from a high concentration runoff than when a more dilute runoff enters. Similarly, the removal efficiency would be greater in response to a small storm than to a long, intense storm. Anticipated ranges are provided in Table 6.3.

Table 6.3.   Anticipated effluent characteristics from a settling basin designed to prevent solids in feedlot runoff from entering a runoff retention basin

Parameter	Performance
Total solids concentration	4,000–10,000 mg/l
Total solids removal	40–80%
	Removal is greatest when high solids concentrations enter the basin such as during snow melt or during the early part of a storm event
Total nitrogen concentration	200–500 mg/l
Total nitrogen removal	20–60%
Organic matter removal	20–50%

Table 6.4.   Anticipated solids removal efficiencies of centrifuges treating dilute manure slurries from various animal species

Parameter	Dairy manure slurry, %	Poultry manure slurry, %	Beef manure slurry, %
Total solids removal	60–70	70–80	60–70
BOD removal	30–40	35–40	25–35
Solids fraction solids content	20–25	25–30	20–30

Note: Appropriate addition of filtering aids can be expected to increase solids removal by up to 10% and to increase the solids concentration in the solids stream by up to 20%.

Centrifuges, although currently less popular for livestock waste handling than settling basins, depend on particle density as the means of separation (Figure 6.2). Centrifuges are more frequently found in industrial applications where the flow rate is nearly constant. Centrifuges are designed to spin the water in the bowl of the centrifuge so the solids experience an effective gravitational force several times greater than would be encountered in a settling basin. This spinning allows a more compact solid fraction to be harvested. Centrifuges are generally designed to handle several hundred gallons per minute, which is beyond what is needed by most livestock operators on a continuing basis. As wastewater processing becomes more advanced, it is reasonable to expect that centrifuges will become more widely used because of their compact size. In considering the use of centrifuges for solids recovery, whether it be from a fresh, untreated wastewater or a solids-laden basin, it is helpful to know there are a number of filtering aids that can be added in trace concentrations to greatly improve centrifuge performance. Among these aids are the well-established lime and ferric chloride as well as more recently devised polymers that help fine solids clump together and more freely release water. Typical performance data for centrifuges are given in Table 6.4.

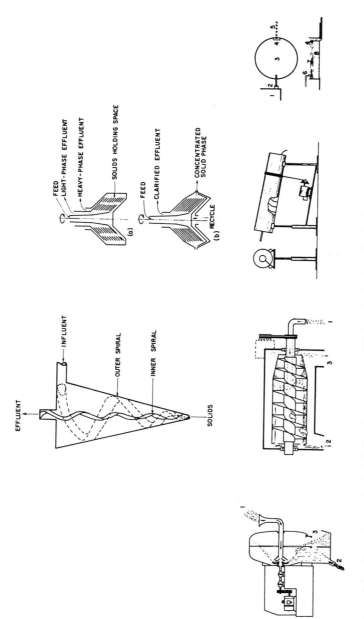

Figure 6.2.   Alternative mechanical devices for separating solids from liquids in hydraulically transported manure.

Separation Based on Particle Size

Screening processes are relatively obvious and can be easily imagined. Unfortunately, when first applied to the separation of solids from manure slurries, standard screens were met by disaster because they quickly clogged and proved to be such an operational headache that most operators removed them and decided that screening was not a useable tool. Fortunately, equipment is available that overcomes many of these early problems so that screening has become a common practice. The most obvious need is to devise a screening process in which the screen does the separation rather than the material that is accumulated on the screen. Passing a complex mixture of particle sizes through a standard woven-wire screen, the screen first removes particles too large to pass through the screen openings but almost immediately this layer of accumulated material forms a layer that also screens subsequent material. Each succeeding layer traps smaller particles and very quickly the screen plugs because there are a large number of bacterial cell-sized particles in manure that are small enough to interfere with the passage of water.

Sloping stationary screen. The sloping stationary screen is the most popular choice for the separation of solids from fresh manure because of its simplicity of operation, lack of moving parts, and generally low cost of operation. For this device, the liquid is distributed across the width of the screen at the top edge where the screen is nearly vertical (Figure 6.3). Solids larger than the open space between the screen strands pass through the fabric. The solids that are retained on the surface are moved down the screen by gravity and the flowing liquid until they finally drop off the lower edge. Note particularly the shape of the individual strands of the screen fabric (Figure 6.3). The nearly triangular shape is important in overcoming one of the traditional screening problems, the solid particles passing partway through the screen then becoming stuck. In this screen design, the most constrictive dimension is at the surface. A solid particle that passes through the surface then falls through the screen. The flow of incoming liquid tends to move accumulated solids down the incline. As solids approach the end, they have additional time to drain. Sloping screens are available with a variety of spacing between the bars. Openings of approximately 0.5 millimeters are common for separating manure solids from dilute slurries. Originally, sloping screen fabric was made of stainless steel that has a long, useful life but is relatively expensive to purchase. More recently, alternate and less expensive material has become available. The designer is therefore required to evaluate expected screen life span versus cost. Typical sloping screen performance is summarized in Table 6.5.

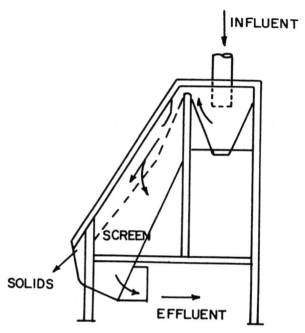

Figure 6.3. Schematic of a sloping stationary screen solid–liquid separation device.

Table 6.5. Typical constituent removal for a sloping screen receiving beef or dairy cow manure slurry with an initial solids concentration between 1 and 2%

Parameter	Range, %
Total solids removal	50–70
Volatile solids removal	50–75
BOD removal	40–60
COD removal	40–60
Organic nitrogen (N) removal	30–50
Ammonia N removal	20–30
Solid phase solids content	13–20

Note: Removal efficiencies for swine and poultry wastes are likely to be less than those for beef and dairy cow slurries because the feed is generally ground more finely.

Sloping screens are not maintenance free even though there are no moving parts and the system is relatively simple. The most frequent need is cleaning to remove bacterial slime growths that accumulate on the fabric, restricting the passage of liquid. Cleaning can be accomplished with a daily scrubbing or by the installation of an automatic brushing mechanism that travels the length of the screen. Some operators spray the screen periodically with a chlorine solution to control bacterial growth. In a few instances, struvite particles have been reported as accumulating on the screen. Struvite is best removed by spraying with boric acid or an alternative mild acid solution.

Vibrating screens. Vibrating screens are the second most popular way to separate manure solids from liquids. The vibrating screen is used in a variety of industrial applications and is a highly flexible device that can be tuned to perform at an optimal level for different waste materials. The essential operation is that the slurry is pumped or flows onto a horizontal screen fabric that can be selected with various sizes of openings. The liquid fraction passes through the openings and the solids remain on top of the fabric. The unit vibrates based on having weights mounted asymmetrically on a rotating shaft supporting the screen unit. The vibrations accomplish two purposes. First, they cause the trapped solid fraction to move across the screen surface to an outlet chute. Second, they cause vertically mounted cylinders immediately below the screen fabric to move relative to the fabric, cutting off any fibrous material that might tend to plug the fabric by stapling.

Separation results from a vibrating screen are approximately the same as those reported in Table 6.5 for the stationary screen. Vibrating screen performance in terms of the fraction of the solids removed is related to the size of openings in the screen fabric. Choice of screen opening size becomes a tradeoff guided by experience; a finer screen opening removes more of the solid material but has less capacity and tends to produce a less dry solid fraction compared with coarse fabric. A coarse fabric allows more material to pass through but generally allows higher flow rates and produces a solid fraction with a lower moisture content compared with fine fabric. A schematic drawing of the vibrating screen is shown in Figure 6.4.

Other filtering devices. There are other filtering systems being used for separating fresh and treated manure solids from liquids, however, they are less common than the stationary slopping screen and the vibrating screen. Rotary vacuum filters have been adapted from the industrial and municipal wastewater field for the separation of anaerobically digested manure solids from the transport liquid. These units are large and are of

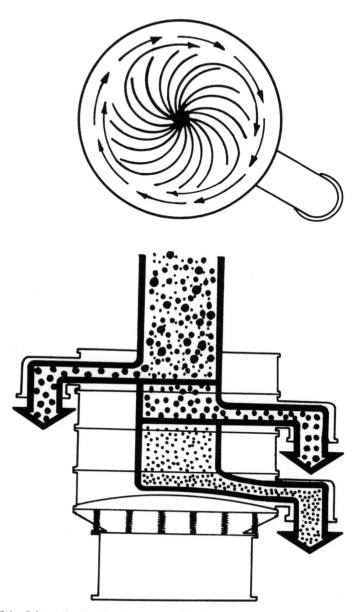

Figure 6.4.   Schematic of a vibrating screen solid–liquid separator.

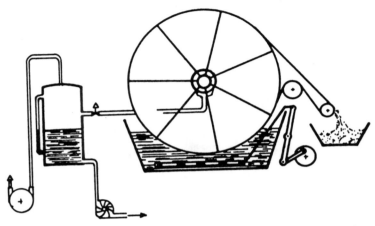

Figure 6.5.   Schematic of a vacuum filter to remove and de-water manure solids from transport liquid.

the general design shown in Figure 6.5. It is typical to add a filtering aid to the slurry before passing it through a rotary vacuum filter to improve the separation efficiency.

Several manufacturers produce solid–liquid separation equipment based on the use of a traditional belt filter. The principal is similar to the vacuum rotary filter but the physical arrangement is very different. One device deposits the manure slurry on a porous belt that subsequently passes through a series of rollers that press the liquid through the belt and allow the solid fraction to be carried through the device to a storage pile. This system is particularly good when a dry solid fraction is needed because the solids have been squeezed as well as filtered.

References

Midwest Plan Service. 1985. Livestock Waste Facilities Handbook, MWPS–18. Ames: Iowa State University, MWPS.

Miner, J.R., A.C. Goh and E.P. Taigainides. 1983. Dewatering anaerobic lagoon sludge using a decanter centrifuge. Transactions of the ASAE. 26(5): 1486–1489.

7

Anaerobic Treatment

Anaerobic decomposition is one of the most common processes in nature and has been extensively used in waste and wastewater treatment for several centuries. New applications and system modifications continue to be adapted making the process either more effective, less expensive, or suited to the particular waste in question and the operation to which it is to be applied. In the simplest situation, anaerobic microorganisms, principally bacteria, decompose most naturally occurring organic material in the absence of free oxygen. If animal manure is placed in a container of almost any size or shape and stored at a temperature above freezing and below 150°F, the solid and dissolved organic material decomposes and yields carbon dioxide, methane, ammonia, and hydrogen sulfide. In the process, a large number of intermediate products are formed and those with a sufficient vapor pressure volatilize, giving rise to the typical odors associated with anaerobic decomposition. Manure and most other natural materials have a bacterial population that over time adapts to the environment of the storage so that bacterial supplementation is generally unnecessary. This process is not restricted to the decomposition of animal manure. Examples of anaerobic decomposition or digestion are common and range from the production and storage of food products such as sauerkraut to the decomposition of forest litter.

The most widely used anaerobic wastewater treatment device is the household septic tank. In many ways, it serves as a highly effective waste treatment device because it is inexpensive to construct, separates the waste material from the living environment, produces an effluent that can be absorbed by the environment without local damage, all with a minimum of maintenance. Fortunately, septic tanks are able to function using a variety of wastewater treatment processes within the same chamber and without seriously interfering with one another. Although the virtues of septic tanks include overall effectiveness and contributions to human convenience and public health, they do not represent the most efficient form of anaerobic digestion, nor do they produce the highest quality of effluent possible using anaerobic decomposition. Process compromises are incorporated into septic tank design to reduce costs and to eliminate operator attention that decreases efficiency.

135

Thus, the design process is one of optimization of the total system not a narrow view of achieving the greatest treatment efficiency. This design understanding is equally important to the designer of livestock and poultry waste handling systems. The goal is to design systems that eliminate or reduce environmental damage to an acceptable level and that do not interfere with the enterprise to the extent to make it economically unjustified or too demanding and time-consuming for the operator.

Anaerobic Digestion Process

Anaerobic digestion is important in nature as a recycling process and in engineered processes as a way to transform low-value products into higher-valued products or to transform materials that are difficult to reuse into forms that are more desirable. Because our concern is livestock waste management, we are interested in the transformation of waste materials into forms that are less damaging to the environment, more conveniently used, or of greater value. Whatever the application, the process is essentially the same. Anaerobic decomposition is complex but has been extensively studied. Humans consider anaerobic decomposition as essentially a breakdown process that converts complex manure constituents into simpler, more easily used end products; to the microorganisms responsible for the process, it is one of survival, growth, and reproduction.

Anaerobic bacteria can survive and grow in environments that are devoid of free oxygen. Humans and other mammals are restricted to life in an oxygen-containing environment. Some bacteria are also obligate aerobes, but others such as those that grow in septic tanks and other anaerobic environments thrive in the absence of oxygen. Aerobes use oxygen as the final proton acceptor in energy production. It is a very efficient form of energy production. Anaerobes are restricted to other proton acceptors such as chemically combined oxygen found in organic acids, ketones, sulfates, and nitrates. This form of energy production is less efficient so they must process more material to achieve similar quantities of energy as produced by aerobes.

Although anaerobic decomposition, sometimes called anaerobic digestion, is a complex process involving numerous individual reactions and a variety of enzyme-excreting bacteria, it is often simplified and described as a two-stage process. The first stage is one of acid formation in which the carbohydrates, proteins, and fats in the manure are attacked by a hearty group of putrefying bacteria. These bacteria are present in the environment, multiply rapidly in response to a food supply, and are highly effective in converting organic material into simpler compounds, including short-chain organic acids. These short-chain organic acids are commonly called volatile acids because they are typically measured using a steam distillation process that collects only those compounds with a relatively high vapor pressure. The second stage of anaerobic decomposition, sometimes called the acid recovery stage, involves a more sensitive group of bacteria, the methane formers, whose role is to metabolize the organic acids, converting them to carbon dioxide and methane. These methane formers are sensitive to low pH, temperature change, and any other inhibiting additives. Thus, it is important that a balance be achieved so acids are converted to methane and carbon dioxide, which leave the system, at about the same rate as they are formed. Design and operator attention are crucial in making this happen.

Conversion of Manure Components into Energy

From a biochemical perspective, manure is a complex mixture of carbohydrates, proteins, and fats. Each of these components can be used as a source of energy and for cell reproduction by anaerobic bacteria.

Carbohydrate decomposition. The initial decomposition of complex carbohydrates such as starch and cellulose occurs via hydrolysis in which extra cellular enzymes excreted by a variety of bacteria break the bonds between the six carbon ring-structured sugars by the addition of water. This process yields a variety of simpler compounds that can then pass through bacterial cell walls for further degradation and energy production. Not all of the linkages are equally easy to hydrolyze, thus certain carbohydrates decompose more easily than others. Cellulose is among the more resistant of the carbohydrates to hydrolysis. Aerobic bacteria are less able to decompose cellulose than anaerobes.

Once the complex carbohydrates have been cleaved into simpler six carbon-containing sugars, or hexoses, they can be metabolized to pyruvic acid in an energy-releasing step. From pyruvic acid, the process can follow a number of alternate pathways that involve the production and subsequent metabolism of a variety of organic acids, alcohols, and ketones. It is important to note at this point that the anaerobic digestion of carbohydrates produces organic acids as a necessary step in the process. It is important from a design perspective that there be a sufficient mix of bacteria in the process so that these acids do not accumulate and cause the pH of the digesting material to become inhibitory to the bacteria performing subsequent decomposition.

Protein, fat, and oil decomposition. Protein is an important ingredient in the ration of livestock and poultry for growth, reproduction, and milk production. As with carbohydrates, the first step in protein decomposition is a hydrolytic process in which the long chains of amino acids that make up the proteins are hydrolyzed into ever shorter amino acid chains. Once individual amino acids have been released, they can penetrate the cell wall where they are used for energy. Among the energy-releasing steps is the deamination in which the ammonia of an amino acid is removed and released. Proteins include sulfur-containing amino acids whose break-down provides the necessary ingredients for complex and highly odor-ous sulfur-containing compounds such as skatole and indole as well as ammonia.

The amount of fat and oil in animal and poultry manure is usually small because of the limited amount included in most feed rations. Be-cause of these low concentrations, fats and oils can be decomposed by the anaerobic bacteria, yielding glycerol and various fatty acids. Both are further processed by anaerobic bacteria in a multistep process leading to methane, carbon dioxide, ammonia, and hydrogen sulfide. During this process, the large number of intermediates formed, depending upon the design of the system, may volatilize.

Factors Influencing Anaerobic Treatment

Anaerobic waste treatment is a biological process that depends on the growth of a variety of anaerobic bacteria. The process is intended to anaerobically decompose manure constituents into simpler, lower molecular weight compounds while converting a major portion of the or-ganic solids in the manure first to a liquid fraction and subsequently to methane, carbon dioxide, ammonia, and hydrogen sulfide. Anaerobic treatment may not go to completion but can be designed to achieve lev-els of organic removal ranging from 50 to more than 90%.

Temperature. Temperature is an important variable in determining the speed of a biological process. Anaerobic digestion is particularly sensitive to temperature. Traditionally, heated municipal sewage sludge digesters have been designed to operate at 35–37°C (95–98°F). This process is called mesophilic (medium temperature) digestion and has been car-ried over to anaerobic digesters for animal manure. Thus, a lot of the ex-isting digesters are operating at this temperature range. Anaerobic di-gestion also can operate at higher temperatures, known as thermophilic digestion, 58–60°C (135–140°F). Thermophilic digestion is more rapid than mesophilic digestion and can therefore be conducted in a smaller tank. The advantage of thermophilic digestion is the smaller tank and slightly greater gas yield; the disadvantage is the additional cost of main-taining the digester and its contents at a high temperature and the addi-tional sensitivity of the digestion process at this temperature. Because the

Table 7.1.   Typical hydraulic retention times for well mixed-
anaerobic digesters operated at various temperatures. retention
times can be reduced in mesophilic and thermophilic digesters if
provisions are included to increase the biological solids (active
anaerobic bacteria) retention time

Temperature	Hydraulic retention time (days)
55–60°C (130–140°F)	10–15
35–38°C (95–100°F)	30
20–25°C (68–75°F)	60–90

biological process is operating at a more rapid rate at this higher temperature, changes in the system are more rapid and are generally regarded as requiring a higher level of management.

Anaerobic digestion also occurs at temperatures below 35°C and is
called psychrophilic (cold loving). Low-temperature digestion such as
might take place in an anaerobic lagoon or septic tank can be equally effective for waste treatment but probably cannot produce gas at a sufficiently rapid or reliable rate to justify gas collection for conversion to
electrical energy. Thus, a mixed anaerobic digestion process requires
greater residence time at lower temperatures than at higher temperatures. Typical hydraulic retention times for well mixed anaerobic digesters at various temperatures are given in Table 7.1.

Mixing. Active anaerobic digestion is dependent upon the available organic constituents being in contact with the microorganisms capable of
conducting the anticipated processes. Thus, it is obvious that mixing
speeds the process compared with a quiescent process where solid materials tend to settle to the bottom and a liquid fraction resides above it. In
this situation, pockets of adverse environmental conditions are likely to
develop that could inhibit effective digestion. Therefore, digester mixing
to speed digestion, maintain a uniform temperature, and avoid digestion
inhibition is an important design consideration.

In contrast to digester mixing, there are a large number of plug flow digesters that have been built recently in Asia that are performing satisfactorily. The typical Asian version of the plug flow digester is a concrete-lined
channel of rectangular cross section and of an appropriate length to provide
a hydraulic retention time of approximately 30 days. There is a heat exchange system to maintain a uniform temperature of 95°F. The trench is fitted with a high-density polypropylene or other gas-impermeable, flexible
cover to collect the biogas. The influent is pumped into one end of the digester and displaces an equivalent amount of effluent from the other end.
Plug flow digesters are reported to operate most effectively at 8 to 12%

solids. At lower solids concentrations, there is a tendency for incoming wastes to channel through the digester, effectively reducing the hydraulic retention time. Some newer designs are being built that have some limited mixing within compartments to obtain benefits of both the completely mixed and plug flow extremes.

Another alternative is multistage digestion in which a part of the digestion occurs in a primary, completely mixed digester from which the partly digested slurry flows into a quiescent second stage for completion of the digestion process. This latter alternative is similar to conventional anaerobic sludge digestion as practiced in most modern municipal wastewater treatment plants.

Nutritional requirements. The growth and survival of anaerobic bacteria certainly requires an adequate supply of the full range of nutrients; however, livestock and poultry manure is a plentiful source of these materials.

Inhibition due to pH and organic acids. Anaerobic digestion is a biological process depending on a variety of anaerobic bacteria to convert the manure constituents from complex carbohydrates, proteins, and fats into methane, carbon dioxide, ammonia, hydrogen sulfide, and water. One of the problems to avoid is having volatile acids, which are normal intermediaries, accumulate to levels that inhibit the relatively sensitive methane-forming bacteria from functioning. There are several potential causes of excess acid formation. If the digestion chamber is loaded intermittently, if the temperature is too low, if it is allowed to vary, or if the organic load is in excess of that which can be accommodated, the methane-forming bacteria are likely to be among the first bacteria inhibited. Once they have reduced their activity, organic acids continue to be formed and accumulate. The extent to which this accumulation depresses pH determines the extent of damage. If the buffer capacity of the digesting material is sufficient to maintain a stable pH, the methane-forming bacteria are likely to respond to the additional food supply with a population explosion. If there is inadequate buffer capacity, pH falls, existing methane-forming bacteria cease to function, and conditions in the digester continue to deteriorate. Thus, pH becomes an important operating variable in anaerobic digestion, but pH is so closely related to other parameters such as the buffering capacity (as measured by the alkalinity) that it is misleading to simply list pH as the inhibiting parameter.

Thus, both the designer and the operator of anaerobic digestion processes need to be aware of pH-measuring procedures and to regularly monitor the digestion process. Digester pH changes can be moderated

Table 7.2   Percentage of total ammonia concentration present as free ammonia at various temperatures and pH values

pH	Temperature = 30°C (86°F)	Temperature = 10°C (50°F)
7	<1	<1
8	10	3
9	50	20
10	85	70

by maintaining a large bacterial population. Thus, lightly loaded digesters or those receiving a dilute feed may be at greater risk of depressed pH than the less dilute counterparts that have a greater buffering capacity on a volumetric basis. If a digester is routinely monitored for pH, alkalinity, and volatile acids, and the temperature is maintained at a stable level, it should be possible to avoid or at least minimize digester upsets that require extensive recovery efforts. If it becomes necessary to take action to stabilize an upset anaerobic digester, the usual approach is to reduce the organic load. Another solution may be to add sufficient lime to increase pH to a level that allows the methane bacteria to function.

Inhibition due to ammonia toxicity. Elevated concentrations of ammonia have been blamed for numerous digester problems over the years. Animal manure certainly poses potential problems in that the nitrogen-to-carbon ratio is higher than in most other wastes. It is important to recognize that ammonia toxicity to anaerobic digestion is also a pH-related issue. It is the free ammonia gas in solution or the dissolved ammonium hydroxide that is most toxic. As long as the pH is less than 8.0, more than 90% of the ammonia is present as the less toxic ammonium form. Above pH 9.0, most of the ammonia is present as ammonium hydroxide or dissolved ammonia and is likely to cause inhibition. Experience suggests that if the free ammonia concentration exceeds 150 milligrams per liter, inhibition is likely. Total ammonia concentrations in excess of 1,000 milligrams per liter should be avoided at any pH. The concentration of free ammonia in a solution can be determined based on the total ammonia concentration and the measured pH with data similar to that found in Table 7.2.

Thus, in most anaerobic digesters treating livestock and poultry waste, the ammonia ion is the major contributor to determination of pH, and pH in turn determines the extent to which the ammonia concentration will or will not be inhibiting to the process. Thus, it is important that the operator monitor both ammonia concentration and pH to prevent ammonia toxicity. Ammonia levels can be controlled by adjusting the solids content of the material being fed into the digester as well as the total nitrogen in the manure.

Other toxic materials. Although ammonia is the most common chemical inhibiting anaerobic digestion, it is not the only one. Sodium, potassium, calcium, or magnesium also may inhibit anaerobic digestion if present in sufficiently high concentrations. Fortunately, such high cation concentrations are unusual in livestock waste treatment units. Certain metals also may inhibit anaerobic digestion. Copper has attracted the most attention because it has been added as a feed ingredient in some areas. Copper has limited solubility in the presence of sulfide at pH values typically found in livestock manure slurries, therefore, problems are unlikely. It is wise, however, to inquire if copper is being fed to animals whose manure is destined for anaerobic digestion and to consider the possibility of inhibition and required mitigating procedures.

Sulfides also can prove toxic to anaerobic digestion, however, examples are unusual in practice. Sulfide concentrations of up to 200 milligrams per liter have been reported as not being inhibitory. Sulfides in excess of this concentration would only be anticipated in a location with very high sulfate-containing water. Thus, an alternate water source would probably be necessary for other reasons.

Biogas

Biogas as produced by the anaerobic digestion of organic material, including livestock and poultry wastes, is a mixture containing primarily methane and carbon dioxide. The gas from a well-operating digester is 50 to 70% methane and the remainder carbon dioxide with trace amounts of ammonia, hydrogen sulfide, and volatile intermediates produced in the digestion process. In addition, the biogas is saturated with water vapor in equilibrium with liquid at the temperature of the digester.

Methane has a heat of combustion of approximately 950 British thermal units per cubic feet, therefore, biogas has a heating value of approximately 500 to 550 British thermal units per cubic foot as it comes from the digester. Depending upon the intended use of the gas, it may prove beneficial to clean it up before use. For example, many people using biogas to drive a converted natural gas engine elect to remove the hydrogen sulfide by passing the gas through a trap of iron gauze. It can be further purified if there is justification; however, for most uses there is little incentive.

It is important to remember that methane cannot be as easily liquefied as propane or butane. The critical temperature for methane is $-82°C$ ($-116°F$) and the critical pressure is more than 5,000 pounds per square inch. Thus, it is impractical to consider biogas as an energy source for mobile equipment or to transport it a significant distance for an off-site use. The most successful operations in terms of biogas use are those that can benefit from the energy on the farm where it was produced or those that negotiate an attractive buy-back contract with their electrical energy supplier.

Anaerobic Digester Design and Operation

Heated and stirred anaerobic digesters have been used in wastewater treatment for many decades and for the past 25 years have been recognized as effective in the breakdown and stabilization of livestock and poultry manure. In addition to providing effective stabilization of the organic matter, anaerobic digestion conserves a greater portion of the nutrients than alternative treatment technologies because the process is conducted in a closed container, thereby avoiding ammonia volatilization. In addition, because anaerobic digestion is conducted in a sealed tank, it is possible to avoid odor release that characterizes anaerobic processes that are conducted in open basins such as anaerobic lagoons. In addition to the treatment and odor control benefits of anaerobic digestion, biogas is produced that can be used on site as a heating fuel or that can be used to power an engine–generator set that produces electrical energy to be sold for off-site use.

In spite of these benefits, anaerobic digesters are currently being used on a small percentage of livestock and poultry farms in the United States. There are several reasons for this relatively low rate of adoption. First, anaerobic digestion requires an engineered system that is more complex and more expensive to construct than alternative manure treatment and storage systems such as anaerobic lagoons and storage tanks. Second, and perhaps the greatest constraint to the implementation of anaerobic digestion technology, is the widespread availability of low-cost fossil fuels during the past three decades. There was an intense interest in anaerobic digesters during the fuel shortages of the early 1970s. Other constraints are summarized in Table 7.3.

Anaerobic digesters have been more widely adopted in other parts of the world than in the United States. Small-scale anaerobic digesters have been widely adopted in Asia and in western Europe where land availability is limited. In Europe, digesters may be accompanied by biogas use on farms as a source of heat and lighting. In Asia, biogas is converted to electrical energy both on the family farm and in larger commercial livestock operations. In both of these areas, petroleum has been more expensive relative to personal income than in the United States. In addition, land is less available for the construction of anaerobic lagoons.

Table 7.3. Benefits and constraints relative to the use of anaerobic digesters in the treatment of livestock and poultry manure in the United States

Benefits	Constraints
1. Anaerobic digestion is conducted in an enclosed tank system that reduces the release of ammonia and other odorous gases.	1. Anaerobic digesters are more complex and more expensive to construct than are alternative waste management devices.
2. Produces a biogas that is 55 to 65% methane, thus it has value as a fuel.	2. Anaerobic digesters are more difficult to operate than alternative waste treatment or storage devices.
3. Produces an effluent in which the nutrients are in a form more easily applied to pasture and cropland. The nutrients are also in a soluble form that facilitates further by-product recovery as a feed ingredient or as a raw material for further processing.	3. There has been a plentiful supply of low-cost fossil fuel available during the past 25 years, thus it has made it difficult to market the biogas at a price that justifies the additional investment.
	4. The effluent from an anaerobic digester is sufficiently rich in organic matter and plant nutrients so it is not acceptable for stream discharge.

Configuration of Anaerobic Digester Systems

There are numerous possible anaerobic digester configurations. For this discussion, three configurations can be identified and additional variations can be projected from these configurations. The first and most common in the United States at this time is the completely mixed anaerobic digester (Figure 7.1). The second would be the plug flow system (Figure 7.2), the preferred design in Taiwan and much of the rest of Asia. There is a third, more sophisticated type of anaerobic digester in which some provision is made to increase the retention time of the active bacteria in the tank while reducing the time of the liquid fraction. This can be accomplished by recirculating organic solids from the digester by solid media packed into the digester or honeycomb-type structures within the digester to which the bacteria cling rather than being washed out with the effluent.

Completely mixed anaerobic digesters. The digester design shown in Figure 7.1 is an adaptation of the standard digester that has been used in many municipal wastewater treatment plants over the past 80 years. One of the keys to the successful operation of a digester of this type is to get the manure into the digester as soon as possible after excretion and with as little additional water as possible. Typical solids content ranges from 3 to 12%. When the solids content is less than 5%, the cost of the tank

Figure 7.1. Enclosed, heated, and mixed anaerobic digester designed for the capture and use of biogas.

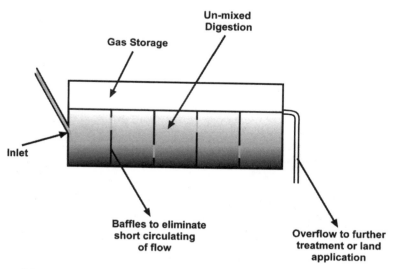

Figure 7.2. Plug-flow anaerobic digester.

becomes a major obstacle to economical design. Tanks have been constructed of coated steel, reinforced concrete, and plastered brick.

Safety Alert

Biogas is a mixture of methane and carbon dioxide with small amounts of other gases such as ammonia and hydrogen sulfide. Whether produced in an anaerobic digester or in a partially filled manure storage tank, biogas is an asphyxiant. No person should be allowed to enter an anaerobic digester to repair plumbing fixtures or any other purpose without being absolutely assured of a source of breathing air.

In addition, biogas when mixed with air, is an explosive mixture. When liquid is to be removed from an anaerobic digester, provisions must be made to be certain a positive pressure is continually maintained in the digester to avoid pulling air into the digester.

Digesters must be fitted with the appropriate flame arresters and other safety devices to avoid a catastrophic explosion. Electrical equipment around digesters should be selected to ensure that electrical sparking does not create a fire or explosion.

Both fixed and floating covers have been used. Floating covers have the advantage that biogas storage can be in the same tank as the digestion and the weight of the cover provides the pressure necessary to move the gas to the point of use. This design may be used to avoid the construction of a separate gas storage structure but the trade-off is that if gas storage is to be in the same tank, a portion of the digestion volume is sacrificed. If a fixed cover design is selected, there needs to be a pressure-regulating device to ensure there is a positive pressure in the digester at all times to avoid pulling air into the digester. Remember that a mixture of biogas and air is explosive.

Various methods have been used to introduce fresh manure to the digester. Remember the digester is under a slight positive pressure so do not anticipate gravity delivery of the influent. It is also important that the manure entering the digester be of the appropriate water content to avoid filling the digester with too dilute a mixture.

Hydraulic detention times for completely mixed anaerobic manure digesters typically range from 10 to 30 days, depending upon the intended operating temperature (Table 7.1). Temperature and mixing ca-

Table 7.4. Anticipated biogas production from anaerobic digestion of manure from 1,000 pounds of livestock and poultry

Animal	Biogas production, cubic feet per 1,000 lb body weight	Heat value, BTU/day	Propane equivalent, gal/day	Potential electrical energy, kW-h/day
Dairy cows	39	20,700	0.22	1.21
Beef cattle	22	11,700	0.13	0.68
Swine	28	16,400	0.18	0.96
Poultry, layers	37	22,700	0.25	1.33
Poultry, broilers	51	30,400	0.33	1.78

Source: Livestock Waste Facilities Handbook, Midwest Plan Service, MWPS-18, 1985.

pabilities are the most important variables to consider in selecting the retention time. There is also a trade-off between operating demands and retention time. As the retention time is shortened, the digestion process becomes more fragile and considerations such as regular feeding, thorough mixing, and temperature stability become increasingly critical. Operator attention needs to be more regular and is more demanding as the design retention time is lowered from the traditional 30 days.

The amount of biogas that is produced is a function of the amount of carbonaceous organic matter that is digested. Most designers base their estimates of biogas production on 8 cubic feet of biogas being produced per pound of digestible volatile solids being fed into the digester. Thus, it is important that the manure be introduced into the digester as soon after excretion as possible. Any decomposition that occurs prior to entry into the digester represents a loss of potential gas production. The potential gas production from various animal species is summarized in Table 7.4. These estimates are based on total manure collection and no decomposition outside the digester so they should be viewed as potential rates and not necessarily rates that can be achieved in a particular installation. These gas production rates represent total gas production and do not recognize the use of a portion for maintaining the digester at its operating temperature or for heating the influent manure. Digester mixing also consumes energy that needs to be considered in estimating the net energy available for use.

There are several alternatives that have been used for mixing anaerobic digesters. The most popular is that shown in Figure 7.1 in which a low-head centrifugal pump is used to pump from the bottom of the digester through a heat exchanger and discharge back into the tank so as to create a mixing action. Earlier digester designs included a mechanical stirring device with a sealed shaft passing through the roof to a drive motor. This design was highly effective when first installed but proved difficult to maintain. Gas mixing is another possibility in which biogas is pumped through a distribution system near the bottom of the tank, creating a rolling action within the tank. The challenge of this system is that like the mechanical stirring devices, the equipment is located within the tank where it is difficult and potentially dangerous to service and repair.

Anaerobic digester effluent. The effluent from a well-operating anaerobic digester is qualitatively as well as quantitatively different from the influent. Most of the nitrogen and phosphorus has been retained. The nitrogen that was initially present as organic nitrogen has been converted to ammonia. The organic matter has been degraded and the volatile solids concentration reduced by 60 to 80%. The tendency of fresh manure to hold water has largely been overcome so the solids from a digester can be dried on a sand bed. Innovative livestock producers are evaluating alternate uses for the digested solids, including incorporation into feed rations. Single-cell protein production also has been considered. Currently, however, most anaerobic digester effluent is discharged into a holding basin where it is stored for later application to crop or pastureland. The advantage of this process is that the holding basin is considerably less odorous than the equivalent anaerobic lagoon. The digester effluent is also amenable to further treatment if land is not available for nutrient use.

Two-Stage Anaerobic Digesters

A popular variation of the completely mixed anaerobic digestion process is the two-stage digestion process. In this variation, two-thirds to three-fourths of the digestion capacity is provided in a completely mixed and heated digester. Displaced digester contents then flow to a second-stage digester that is smaller and neither stirred nor heated. The material entering this tank is allowed to settle and settled material is pumped back to the inlet of the primary digester. This option has the benefit of increasing the active bacterial population in the primary digester and enhancing the quality of the effluent from the second-stage digester by reducing the solids concentration. The disadvantage is that a second tank is required.

There has been limited incentive in the past to obtain the benefit of this more elaborate process, however, as land areas for lagoons become more difficult and land application sites less available, this alternative is likely to become more attractive. Municipalities have tended to build two-stage digesters as a way to reduce the amount of sludge requiring land application because of hauling costs and the need to produce an effluent meeting more strict water quality criteria. Similar trends are likely in the treatment of livestock waste.

Fixed Film Digesters

Another alternative having many of the advantages of the two-stage digester is the fixed film digester in which a solid media is packed into the digester, providing surfaces for anaerobic bacteria to reside. The concept is to selectively increase the retention time of the active anaerobic bacteria while not increasing the volumetric capacity of the digester. Thus, like volatile solids recycling in a two-stage digester, this concept is one of attempting to increase the efficiency of anaerobic digestion while controlling construction costs. In many respects, the move towards two-stage anaerobic digestion with recycling of the solids and incorporation of fixed films or media is suggestive of the aerobic counterparts, activated sludge and trickling filters. Just as these techniques have allowed more efficient aerobic treatment, the anaerobic versions are suggestive of higher rates of digestion, hence either improved effluent quality, greater biogas yields per unit volume, or smaller digesters for the same waste load. Fixed film digesters operating at thermophilic temperatures are allowing the design and operation of anaerobic digesters at lower hydraulic retention times than was previously considered possible.

Plug Flow Digesters

Plug flow digesters are typically rectangular in cross section and have length-to-width ratios of approximately 10:1. One standard design is to install a flexible cover to the sidewalls of the tank and allow gas pressure to support the cover and develop sufficient gas pressure to force the gas to the point of use. Figure 7.2 is a schematic of such a digester.

The plug flow concept is that there is an overall advantage to a digestion process that reduces or eliminates the possibility that a portion of the fresh input will be discharged on the day it first enters. Certainly, the completely mixed digester does discharge a portion of the material after only a brief period in the digester. The hydraulic residence time of a completely mixed digester is an average value, whereas, theoretically, the plug flow digester would ensure that all of the material would remain in the digester for the full digestion period. Neither of the digesters operates at the extreme. There is some mixing in operating plug flow digesters and there is less than complete mixing in the completely mixed designs.

Usually, the design decision is based on the size of operation. Plug flow digesters tend to be built to serve smaller operations in which mixing and heating equipment would add significant additional cost to the digester. Thus, if a digester is to be built in a warm climate to serve an operation of fewer than 1,000 pigs, a plug flow digester is likely to be selected.

Under similar circumstances, a completely mixed digester would be more likely if the enterprise had 10,000 or more pigs.

In planning for a plug flow digester, it is important that the solids content of the influent be in the range of 8 to 12% solids to increase the likelihood of achieving plug flow. At lower solids concentration, there is likely to be sufficient sedimentation in the digester that the liquid fraction moves through at a different rate than the solids and eventually, the solids settle to the bottom, allowing liquids to pass through the digester in far less than the design retention time.

Anaerobic Lagoon Design and Operation

Anaerobic lagoons first appeared on the livestock waste management scene around 1965. They were an outgrowth of the then-popular facultative waste stabilization ponds that were being widely constructed to provide sewage treatment to a large number of small communities that had previously been unable to have municipal sewers due to the construction and operating cost of conventional sewage treatment plants. As frequently happens, the direct transfer of municipal waste stabilization pond technology to the treatment of animal manure, with its much higher organic matter concentration, was an immediate disaster on many farms. Rather than having built a waste stabilization pond with essentially an odor-free water surface, these manure lagoons were essentially open storage units in which the manure became acidified and emitted odors that were universally unacceptable. They also quickly filled with manure solids that were acid-preserved and showed little tendency towards liquefaction. This experience lead to a prompt understanding there was a dramatic difference between municipal sewage and livestock manure and that if they were to function, lagoons would need to be designed for a specific waste load.

After the less-than-glamorous beginning of lagoons, designers began to understand the anaerobic processes that are necessary in lagoons and to build anaerobic lagoons that met the needs of many livestock and poultry producers. Successful lagoon design requires an understanding of both the process and the local situation in which the facility is to be located. The advantages of anaerobic lagoons are that they can be designed to provide both a storage and treatment function within the same facility and that this facility operates at a minimum of cost compared with other available options. Unfortunately, lagoons have developed a bad reputation in some areas either because they were installed in an inappropriate location, were poorly designed for their location, or were not operated in a manner that permitted them to function satisfactorily.

Three major complaints have arisen regarding lagoons. First and most common is that the odor of lagoons is unacceptable. Second, they have a tendency toward dike failure or overflow, which could discharge a large volume of high-strength organic waste into a waterway. Third, lagoons are of concern due to leakage of high concentrations of soluble nitrogen that can be converted to nitrate, which causes health problems among a select population. This latter groundwater pollution problem may be due to leakage through the lagoon bottom or dikes or may be attributed to overapplication of the lagoon contents to crop or pastureland. Recently, concerns have developed regarding ammonia volatilization from lagoons that is redeposited by rainfall to environmentally sensitive watersheds and coastal regions that require a high level of water quality for primary production and protection of fin- and shellfish. There are technological solutions to each of these problems, but these solutions have not been applied with sufficient rigor to avoid lagoons being regarded as environmental threats in many areas.

Standard Anaerobic Storage Lagoon

Anaerobic lagoons were first built to receive the liquid manure flushed or scraped from livestock confinement buildings and to retain that manure until such time as the manager could beneficially apply it to crop- or pastureland. Experience also indicated that the manure changed in quality during the time it was in the lagoon. Most importantly, the quantity of solids decreased significantly so the operator of a properly functioning lagoon was able to remove lagoon contents with an irrigation pump and there was only a slow rate of solids accumulation on the lagoon bottom. This can be regarded as an ideal situation. First, pumping lagoon liquid is a much less demanding job than hauling manure. Second, having the manure nutrients available when they are useful is an advantage over having to haul manure onto frozen ground or when it is otherwise of no value to the cropland. And third, the noticed decrease in nitrogen and phosphorus quantities was considered relatively unimportant and in some cases an advantage because of limited land availability for manure use. Figure 7.3 is a schematic presentation of this process. This somewhat idealized solution makes an appropriate base for proper lagoon design, construction, operation, and maintenance.

The anaerobic digestion process in an anaerobic lagoon is essentially the same as that discussed for digesters, except that without temperature control, the rate depends upon the climate of the lagoon location. Thus, in colder climates, anaerobic decomposition essentially ceases during a part of the cold season. This means that organic material that flows into the lagoon during this time is stored but very little treatment other than

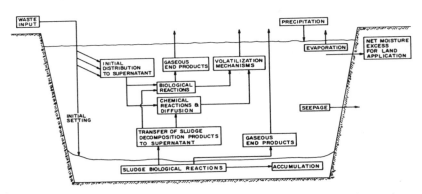

Figure 7.3.   Schematic representation of physical and biological processes that are important to the functioning of an anaerobic storage lagoon.

sedimentation occurs. Both soluble and particulate organic material accumulates. The particulates settle to the bottom and the sludge layer increases in depth. The liquid fraction increases in organic matter concentration, waiting for the time when the water temperature increases sufficiently for bacterial activity to resume or increase. When the temperature of the water reaches approximately 5°C (40°F), signs of biological activity become evident. As the temperature increases, activity becomes more intense until the backlog of accumulated organic material is processed. It is during this warming period that anaerobic lagoons are most likely to evidence odor problems because of what is effectively a more heavily loaded situation in which the routine waste load is supplemented by the material that was stored during the cold period.

Sizing a lagoon. The most critical decision in lagoon design has to do with the volume. There are four functions for which lagoon volume is important so it is appropriate to allow capacity for these functions. Sometimes, the volumetric capacity needs are such that the total volume can be less than the total of individual needs, however, more frequently, all of the requirements are additive. Table 7.5 lists the components of a lagoon function that require volumetric capacity.

TREATMENT VOLUME

The treatment volume of a lagoon is most reasonably based on the amount of organic material entering. Traditionally, the treatment volume was based on the weight or mass of volatile solids being discharged into the lagoon on a daily basis. More recently, designers have tended to

Table 7.5. Summary of lagoon functions for which volumetric capacity needs to be included during the design process

Function	Basis of design	Hazard of undersizing
Biological stabilization of the organic component of the entering waste.	Typically based on the weight or mass of volatile solids being discharged into the lagoon on a daily basis. Alternatively can be based on the number, size, and animal species being served.	Lagoons with inadequate treatment volume typically accumulate sludge faster than anticipated, tend to have more severe odors during the spring warming time, have increased ammonia volatilization, and tend to release more odor at the time lagoon contents are removed for land application.
Storage of biologically stable sludge that remains after anaerobic decomposition of the solid fraction has gone to completion.	Theory suggests that it is possible to predict long-term sludge accumulation. In practice, however, if adequate treatment capacity is provided, sludge accumulations in excess of 2 feet are unlikely.	When sludge storage is inadequate and relatively inactive sludge accumulations invade the treatment volume, causing a reduced degree of treatment with the associated odor problems outlined above.
Storage of treated liquid effluent.	The volume allocated to effluent storage can be calculated based on the net daily inflow volume and the desired frequency of effluent removal.	Inadequate effluent storage capacity causes a lagoon to fill with liquid sooner than planned and in severe cases can lead to uncontrolled discharges or in more monitored situations, forces effluent removal at times less desirable or less convenient than planned.
Storage of runoff and rainfall striking the lagoon surface.	All lagoons gain volume as a result of heavy rainfall. Those that are designed to receive the runoff from manure-covered surfaces or other areas receive more. It is important to include an appropriate volume to safely hold this runoff. Many of the disastrous lagoon failures have been caused by inadequate runoff storage capacity.	When lagoons receive rainfall or runoff in excess of their capacity to retain it, discharge is likely. The alternative is to land-apply lagoon contents as an emergency procedure. Typically, soils are too wet to appropriately receive effluent at these times, however, this may be the least hazardous alternative in an emergency.

Table 7.6.   Anaerobic lagoon treatment volumes (cubic feet per pound of animal contributing manure) based on the three climatic zones of Missouri for different animal species

Animal	Northern Missouri, cu ft/lb	Central Missouri, cu ft/lb	Southern Missouri, cu ft/lb
Dairy cows	2.8	2.3	2.0
Beef cattle	1.8	1.5	1.3
Calves on liquid concentrate	1.3	1.1	0.9
Calves on roughage	2.7	2.3	1.9
Swine, finishing	1.4	1.1	1.0
Swine, sows and litter	1.8	1.5	1.3
Swine, boars	0.7	0.6	0.5
Poultry, layers	2.7	2.3	2.0
Poultry, broilers	5.6	4.7	4.0
Sheep	3.5	2.9	2.5
Horses	2.4	2.0	1.7

Source: Adapted from Missouri Approach to Animal Waste Management. Planning and Design Guidelines. Manual 115, University of Missouri Cooperative Extension Service, 1979.

make these calculations based on the number, size, and animal species being served. For example, Missouri distributes an animal waste treatment facility design guide that specifies lagoon volume based on the weight of animals served. Separate values are suggested for the northern, central, and southern regions of the state (Table 7.6). These regional differences reflect the severity of winter weather in each of the three regions. On a broad basis, appropriate lagoon size recommendations are even more variable. The American Society of Agricultural Engineers suggests a procedure in which the United States is divided into seven climatic zones for lagoon design (Figure 7.4). Table 7.7 summarizes suggested lagoon volumes based upon their approach.

Notice in Table 7.6 that the suggested lagoon volumes differ for animal species reflecting differences in their digestive functions and differences in the nature of the rations typically fed. For example, boars require a smaller active lagoon volume per pound body weight than finishing animals. This largely reflects that boars are fed less per pound because they are no longer growing. Dairy cows require larger lagoon volumes than beef cattle, again reflecting the higher feeding rate for a lactating animal. Note that cattle on roughage require approximately twice the lagoon volume as those on concentrate because of the less digestible feed material, hence the greater amount of manure to be anticipated.

Comparison of Tables 7.6 and 7.7 is a good way to understand that lagoon design is not an exact science but one that calls upon the designer to consider the lagoon operating experience of the area involved, the risk of having an undersized lagoon, and the cost of having a larger lagoon.

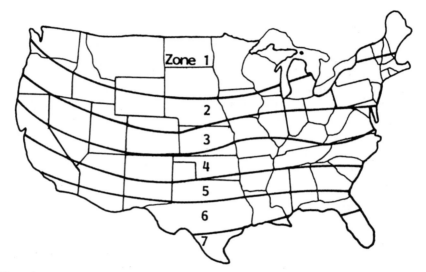

Figure 7.4. Map of the United States showing various climatic zones used as a basis for the volume necessary to treat swine waste (cubic feet of lagoon volume per pound of swine). Warmer weather leads to smaller lagoon volumes. Zone 1, 3.0; zone 2, 2.5; zone 3, 2.0; zone 4, 1.9; zone 5, 1.8; Zone 6, 1.6; zone 7, 1.4.

Table 7.7. Lagoon volumes for various animal species and climatic zones based on animal weight contributing (cubic feet per pound of animal contributing)

Animal	Zone 1	Zone 2	Zone 3	Zone 4	Zone 5	Zone 6	Zone 7
Dairy cow	5.0	4.2	3.8	3.6	3.4	3.2	3.0
Beef cattle	3.2	3.0	2.8	2.6	2.4	2.2	2.0
Sow and litter	3.7	3.2	2.8	2.6	2.4	2.2	2.0
Baby pig	2.8	2.4	2.1	2.0	1.8	1.6	1.5
Nursery pig	2.7	2.4	2.1	1.9	1.8	1.6	1.5
Finishing pig	2.8	2.4	2.1	1.9	1.7	1.6	1.5
Gestating sow	1.4	1.3	1.1	1.0	0.9	0.8	0.7
Boar	1.4	1.2	1.1	1.0	0.9	0.8	0.8
Poultry layers	5.4	5.0	4.6	4.2	3.8	3.5	3.2
Poultry fryers	6.0	5.6	5.2	4.8	4.6	4.4	4.2

Note: Climatic zones based upon severity and duration of winter weather. Zone 1 recommendations are appropriate for northern U.S. conditions; zone 7 volume recommendations would be applicable to southern Florida or other areas where the average temperature during the cold season is seldom below 40°F and summer temperatures are frequently above 90°F.

Typically, experience has been that larger lagoons provide a slightly improved effluent quality based on soluble biochemical oxygen demand (BOD) or volatile solids. Larger lagoons also tend to emit less odor during the spring when the lagoon is recovering from wintertime inactivity. The Missouri guidelines are consistently suggesting a lower lagoon volume than the American Society of Agricultural Engineers' standards. This also may reflect that the Missouri Department of Natural Resources has more specific responsibility to control water pollution than to minimize odor release.

Lagoons with inadequate treatment volume typically accumulate sludge faster than anticipated, tend to have more severe odors during the spring warming time, and tend to release more odor at the time lagoon contents are removed for land application.

SLUDGE STORAGE VOLUME

Anaerobic lagoons do not convert all of the entering solid material into a liquid that can be removed with liquid handling equipment. These accumulated solids whether they are undigested organics or inorganic soil particles tend to reduce the lagoon volume. Theory suggests that it is possible to predict long-term sludge accumulation. In practice, however, if adequate treatment capacity is provided, sludge accumulations in excess of 2 feet are unlikely. The amount of lagoon volume allocated to long-term sludge storage reflects the judgement of the designer as well as the needs of the operator. Typically, lagoons are designed allowing from zero to 1 foot of sludge storage on the bottom. If the treatment volume has been selected conservatively, there is little danger of the lack of sludge storage creating a problem. If, however, the treatment volume was selected based on minimizing the lagoon volume, sludge storage may be critical.

When sludge storage is inadequate and relatively inactive sludge accumulations invade the treatment volume, causing a reduced degree of treatment and the above-mentioned associated odor problems. Unfortunately, this problem tends to be self-supporting. Once inactive sludge invades the treatment volume, treatment becomes less effective, hence the rate of sludge accumulation increases. This chain of events eventually leads to a dead or stuck anaerobic lagoon in which the entire volume becomes filled with manure solids at various stages of decomposition. The lagoon begins to smell bad more frequently and eventually needs to be re-excavated.

STORAGE OF TREATED EFFLUENT

The volume allocated to effluent storage can be calculated based on the net daily inflow volume and the desired frequency of effluent removal. For example, if it was planned to flush a building with 2,000 gallons of water daily and the estimated manure volume was an additional 500 gallons, it would be necessary to have storage capacity for that additional 2,500 gallons each day. If the goal was 180 days of storage, this suggests an effluent storage capacity of 450,000 gallons. One of the temptations to be avoided at this point is to use a portion of the treatment volume for effluent storage. Some operators have managed their lagoons by pumping the level down below the treatment level to create additional storage. This approach essentially creates an overloaded lagoon and contributes to elevated odor problems. Inadequate effluent storage capacity causes a lagoon to fill with liquid sooner than planned, and in severe cases, leads to uncontrolled discharges or in more monitored situations, forces effluent removal at times less desirable or less convenient than planned.

STORAGE OF RUNOFF AND RAINFALL STRIKING THE LAGOON SURFACE

All lagoons gain volume from heavy rainfall. Lagoons designed to receive the runoff from manure-covered surfaces or other areas will gain even more volume. It is important to include an appropriate volume to safely hold this runoff. Many of the disastrous lagoon failures have been caused by inadequate runoff storage capacity. When lagoons receive runoff in excess of their capacity to retain it, discharge is inevitable. The alternative is to land-apply lagoon contents as an emergency procedure. Typically, soils are too wet to appropriately receive effluent at these times, however, this alternative may be the least hazardous in an emergency.

Although no one designing an anaerobic lagoon to store and treat livestock manure plans that the lagoon could discharge, good design makes provision for this possibility. Thus, lagoons should be provided with an emergency overflow that is located in a portion of the dike that results in the least possible damage to the environment. From an operating perspective, it is important to have a plan for what to do if it becomes evident that an overflow is going to occur. It is preferable to remove lagoon contents in a controlled manner rather than to allow a dike to be overtopped and risk the possibility of erosion continuing to cut into the structure, resulting in a massive spill. It is also preferable to apply lagoon contents to crop or pastureland as an alternative to stream discharge. This latter point is true even if it means that wastewater is applied at times when the nutrients cannot be used or even at times when a portion of the applied material runs off. Thus, again the concept of minimizing damage, especially during high-flow periods from chronic or catastrophic rainfall, is critical when it becomes obvious there is no entirely safe option.

Lagoon design details. Although volume is the most critical factor in the design of an anaerobic lagoon, features such as inlets, dike slopes, and other construction details are important.

LAGOON SEEPAGE

Of paramount importance and frequently the subject of regulatory concern is lagoon bottom and dike permeability. Lagoons should not leak because the contents are typically high in ammonia nitrogen. If the bottom or sidewalls allow seepage, the liquid is likely to continue to the groundwater. As soon as unsaturated conditions are encountered, the ammonia is oxidized to nitrate and moves with the water. Soils can be tested to determine the permeability that can be achieved with proper moisture control and compaction during the construction process. State regulatory agencies seem to have settled on a maximum allowable infiltration rate of 10^{-7} centimeters per second. This value is equivalent to approximately 1 centimeter per day, which is not an inconsequential nitrogen loss. The situation is further alleviated, however, by the observation that if the initial infiltration rate achieves this criterion, further sealing occurs due to the development of an anaerobic bacterial slime at the soil–water interface. Thus, it is most critical that in lagoons constructed of native materials that initial sealing is successful.

Field experience is that if the soil in an area selected for a lagoon is underlain with clay material, this material can typically be sufficiently compacted with conventional construction equipment to achieve the permeability requirements. If the site is not underlain with clay adequate to achieve the permeability requirements, one option is to import clay from off-site in sufficient quantities to achieve a 1-foot layer that can be compacted sufficiently to meet the required criteria. A second option is to use a lagoon lining material that is generally a rubber or synthetic material that can be placed inside the lagoon to provide a watertight seal. The hazard of this approach is that during construction or in operation, the lining material could be torn or punctured. This hazard is particularly troublesome if it is necessary to enter the lagoon at some future time to mechanically remove accumulated sludge. In selecting a lagoon lining material, life expectancy and resistance to ultraviolet radiation are important criteria. A third option is that a particular site may be unsuitable for the construction of a lagoon due to being underlain with fractured limestone, sand lenses, large gravel, or other material that raises a leakage concern. In this case, selecting an alternate site is the most appropriate decision.

LAGOON DEPTH

It is generally advantageous to construct anaerobic lagoons at least 10 feet in depth. Depths of 20 feet or more also have been built and have the advantage of achieving the necessary volume with a minimum of surface area for odor release. The important considerations are the soil conditions under the site and the presence of groundwater. Lagoons should not extend into the annual high groundwater table. Most designers prefer to have the lagoon bottom at least 2 feet above the groundwater to avoid contamination. Permitting agencies probably also insist that the lagoon bottom be above the maximum groundwater level. Assuming adequate soil conditions and no groundwater intrusion threats, deep lagoons are desired. Not only is the surface area minimized but also a greater degree of temperature stability is achieved, gas bubbles rise through the deeper liquid to achieve a greater degree of mixing, and the lagoon is more appropriately sized for the addition of a cover or surface aerator if either of these options is selected.

RUNOFF EXCLUSION

Lagoons are designed to provide biological treatment and effluent storage, hence it is important to exclude any water that does not require treatment. Runoff from adjacent areas should be diverted around the lagoon by appropriate dikes, diversions, and surface grading. The top of the dikes should be sufficiently wide to allow any required vehicle access and as a safeguard against dike failure due to erosion. The minimum dike width is 10 feet. It is also appropriate to consider space to turn around, if needed.

LAGOON SHAPE

There is nothing critical or magical about the shape of a lagoon in terms of the biological process. Shape is important however in avoiding solids accumulation in a corner or appendage. The most common shape is rectangular with a length-to-width ratio of 3 or less. Square, round, and freeform lagoons have all proven satisfactory.

DIKE SLOPE

One of the threats to the integrity of a lagoon structure is erosion damage to the dikes. Thus, it is desirable to construct dikes that can be seeded with a grass clover and any vegetation on the dikes can be mowed. Typically, dike slopes are 3 horizontal to 1 vertical, which allows driving a tractor along the dikes for maintenance purposes. Such slopes are sufficiently flat to avoid severe gully formation. Where space limitations require more severe slopes, some form of erosion control such as rock riprap is generally added.

LAGOON INLETS

Both abovewater and submerged inlets have been used satisfactorily for lagoons and each has its advantages. If submerged inlets are used, remember that adequate hydraulic head, 1-foot minimum, needs to be provided above the maximum water level to transport liquid manure into the lagoon. It is easier to support submerged inlet lines than abovewater inlets. The abovewater inlets are easier to inspect to confirm they are intact.

Most lagoon designers prefer to have the inlets discharge near the center of the lagoon to avoid solids accumulating near the bank. This approach may not be practical in a large lagoon, where discharging at least 100 feet from the toe of the dike is probably sufficient. There is some advantage to having multiple inlets to a lagoon larger than 1 acre in surface area to better distribute the solids. Better solids distribution tends to reduce odor production during the spring when the lagoon water warms and anaerobic decomposition is most rapid. Whatever the inlet design, it is important to have easy access to the inlet for cleaning when plugging occurs. One convenient way to accomplish this cleaning is by having an opening or cleanout on the dike where the inlet line crosses.

The size of inlet lines is based on the maximum anticipated flow rate with the additional provision that a 4-inch pipe is the minimum size acceptable to avoid frequent plugging. It is also appropriate to design the slope of the inlet line sufficiently steep to avoid solids settling in the line.

Ramp inlets are sometimes used if manure is to be scraped into a lagoon. The most obvious disadvantage of a ramp inlet is that manure solids and bedding may accumulate near the ramp. It is important both for odor and insect control that the manure is pushed into the lagoon sufficiently far so that solids do not protrude above the water surface. A specially constructed ramp that drops the manure into the lagoon at a deep area is helpful in this respect.

Lagoon start-up. Anaerobic lagoons are biological treatment devices that depend on an active population of anaerobic microorganisms to convert fresh waste into more stable end products. Thus, in the initial establishment of the lagoon, attention to some logical protocol can reduce odors and lead to an effluent of improved quality. Because lagoon loading is an important parameter that influences operation, filling the lagoon up to the design treatment volume avoids an initially overloaded condition. Designers frequently specify that sufficient water be placed in the lagoon to achieve this volume before animals are placed in the facility. This initial filling of the lagoon can come from the water source being used for animal consumption or it may come from an alternate source. Some operators have diverted roof runoff into the lagoon during the initial filling stage. This is an appropriate water source but requires that this diversion be only a temporary measure. Similarly, runoff from surrounding land areas has sometimes been used to provide the water for an initial fill.

The time of starting a lagoon is also important because if a lagoon begins receiving manure in the fall or winter, the material undergoes little anaerobic decomposition until the following warm season. This aspect of lagoon start-up is particularly important in regions in which winter temperatures can be expected to fall below freezing. Lagoons are best started in late spring or early summer when there is the greatest opportunity for a balanced bacterial population to be established.

Some designers have suggested that lagoons be seeded by filling up to 10% of the lagoon volume with sludge from a well-functioning established lagoon in the same area. Experience indicates, however, that if the lagoon is started during an appropriate time of the year, and the lagoon has had sufficient water added prior to introducing the manure, seeding may be unnecessary. Thus, this approach offers another opportunity to balance the benefits of a lower odor release during the initial operation with the additional cost of seeding the lagoon with an established bacterial flora.

Lagoon operation. One of the advantages of a conventional anaerobic lagoon is that it does not require an external power supply other than that to remove liquid for cropland application or for recycling. This advantage, however, does not eliminate management attention. Properly maintained lagoons generally have less odor, produce a more stabilized effluent, breed fewer insects (mosquitoes and flies), and are less likely to develop leaks in the dike.

WATER LEVEL MANAGEMENT

Lagoons are designed to operate with the surface water level between the upper level of the treatment volume and the level that marks the lower level of the stormwater runoff and rainfall storage area. Frequently, a lagoon has a marker post or series of posts installed to help the operator identify these levels. The lagoon level can be monitored to determine the rate at which the level is rising or falling to guide the operator in planning land application procedures. If the lagoon water level is allowed to rise above this level, the operation is in danger of having an illegal discharge if a heavy rainstorm occurs. More dramatic, however, is the possibility that a severe storm could occur that would overflow the dike and erode a sufficient portion of the dike to cause a lagoon failure. Most lagoon disasters reported in the past have been the result of heavy or repeated rain falling on lagoons that were already full, causing the dikes to overtop. This overtopping of an already saturated soil causes an erosion gully to form and it continues to grow until most or all of the lagoon contents are lost.

The alternate water level management error is to pump the lagoon water level too low either in the fall or spring to increase the subsequent storage capacity. Whenever the lagoon is pumped below the design treatment volume level, an overloaded condition is established that leads to an overloaded system from a biological perspective. Overloaded lagoons emit more intense and generally more objectionable odors than those that are properly loaded.

The most effective tool available to the lagoon manager in controlling the water level is application of effluent to crop or pastureland. The amount of effluent that can be applied during an irrigation event and the rate of the application depends upon the crop, the equipment available, the season, and a variety of soil and climatic variables (see Chapter 10). It is important that these constraints be satisfied. In addition, there are advantages for dike permanence to pumping more frequently rather than annually to avoid having a large area of unprotected dike exposed to subsequent rainfall and associated erosion. Multiple pumpings are preferred over annual pumpings. Spring pumping has the potential of reducing lagoon odors due to the heavy load that becomes available as the lagoon warms; however, pumping at this time of year is likely to yield a material of higher nutrient content. This higher nutrient content is an advantage if the nutrients are valuable for crop production. It is a disadvantage if land for nutrient use is a limiting feature of the overall enterprise.

DIKE MAINTENANCE

Maintaining lagoon dike integrity is an ongoing concern. Larger lagoons are of even greater concern. The establishment of a dense, protective grass cover is the first step in dike protection. This cover reduces erosion damage to the dike and rewards the manager by requiring less repair. Dikes should be inspected for erosion damage and for infestations of burrowing animals on a regular basis. Lagoons of 1 acre or less should be inspected after each major rainfall event and at least weekly when no rainfall occurs. Larger lagoons should be inspected more frequently. If evidence of burrowing animals is discovered, immediate steps for their eradication are necessary. Depending upon the animal involved trapping, poisoning, or other steps may be most appropriate. Whatever procedure is selected, it is important that the damage is repaired promptly and that further damage is avoided. Dikes riddled with animal burrows are particular hazards and can result in major failures.

Emerging vegetation at the water line of a lagoon provides an ideal location for fly and mosquito breeding and larval harborage. Because weeds are most likely to grow in these areas, particularly if the lagoon level varies, periodic mowing is important. Depending upon the local climate and the dike condition, mowing frequency may vary from one to six times during the growing season.

Depending upon the animal species involved, the ration being fed, and whether there is a solid–liquid separator ahead of the lagoon, there may or may not be a surface scum layer formed on the lagoon. Scum layers that form a dry crust can be effective in reducing odor emissions, however, less intense scum formation can provide insect breeding, harborage, and depending upon the nature of the floating material can contribute to odors. Dead animals, afterbirth, and other similar materials should not be allowed to enter a lagoon both in terms of visual appearance and their tendency to contribute to odors and insect problems.

Variations on Anaerobic Lagoons

In addition to the "standard" anaerobic lagoon described above, a number of variations or lagoon retrofits are being constructed to fulfill specific needs. These variations include solids removal prior to the lagoon construction, lagoon partitioning with a high level of aeration in the inlet zone for reduction of odor and ammonia volatilization, controlled aeration for partial odor control, and covered lagoons.

Runoff retention basin. Runoff retention basins are constructed to receive the runoff from cattle feedlots and other unroofed confinement facilities and are designed to hold the anticipated runoff from a design storm plus additional storage to provide whatever degree of flexibility is considered appropriate. In many of the states with large numbers of feedlots, runoff retention basins are designed to be preceded by a solids settling facility. In these cases, it is typical to design the runoff retention basin with capacity for the 24-hour, 25-year storm and to plan that the basin be emptied as soon as the crop or pastureland has dried sufficiently to accept the accumulated wastewater. The design of runoff retention basins was discussed in Chapter 5.

An alternative that is common in areas where wastewater application is either unacceptable or economically unwise for a portion of the year is to design runoff retention basins with sufficient additional capacity to store the runoff. In the northern part of the Great Plains, for example, application to land would be inappropriate during the extended periods of frozen ground. In these areas, storage for winter runoff plus a design storm would be required.

Earthen manure storage basin. Earthen basins to store manure and flush-water for the desired period between land applications have been designed in several states. These basins also have sufficient additional capacity to store the anticipated runoff from contributing surfaces plus the runoff from the design storm. In these situations, typically dairies with unroofed holding areas, the normal dry weather waste flow (often referred to as "process water") is the base for the volume calculation and is supplemented by additional volume for storage of rainfall-driven flows. The liquids in the storage basin undergo anaerobic decomposition to the extent possible but there is no capacity allowed for this function. As a result, they have the potential to operate more like liquid manure storages than lagoons. Storages of this type exist throughout the central United States from Minnesota to Texas. Earthen storages may be preceded by a solid–liquid separator to facilitate removing the stored material. Alternatively, there may be agitation equipment specified to suspend the manure solids during the emptying process. Manure may be flushed or scraped to an earthen storage basin depending upon the type of housing provided.

The quality of material removed from an earthen storage basin depends upon the manure handling system as well as any changes in quality that occur during storage.

Covered anaerobic lagoons. Covered anaerobic lagoons have been built partly in response to the interest in biogas recovery and partly in response to the need to achieve a higher degree of odor control than provided by conventional anaerobic lagoons. Gastight covers manufactured from a variety of natural rubber and synthetic materials reinforced to provide durability are available. These covers can be applied to an existing lagoon or may be included in the initial design consideration. When covers are included in the initial lagoon design, loading rates are typically higher than would otherwise be the case. Figure 7.5 shows one of the standard approaches to the installation of an impermeable cover over an anaerobic lagoon.

Lagoon covers reduce odor release in two ways. First, they allow the gaseous products of anaerobic decomposition to be contained. Second, they reduce the escape of odorous compounds from the liquid surface by preventing wind from continually sweeping away the released gases, thereby greatly reducing the effective vapor pressure difference between the liquid and gas phases.

Figure 7.5.   Anaerobic storage lagoon fitted with an impermeable cover for odor reduction and the potential for biogas recovery.

If biogas is to be collected and burned as a fuel or to drive a genera-tor, gas-handling equipment similar to that described for an anaerobic digester is necessary. Biogas production from a covered anaerobic la-goon is highly temperature-dependent and can be expected to fall to a low rate in areas where temperatures are below freezing for extended pe-riods. Thus, for biogas from a covered lagoon to be attractive, there must be an economically viable use for the gas during the period when it is available. Typically, the operator of such a facility has a contract that al-lows sale of electrical energy to the local utility or alternatively allows the livestock producer to use the energy returned to the system to offset his or her use at another time.

It is also possible to use a covered anaerobic lagoon without using the biogas as an energy source. One option is to burn the captured gas in a waste gas burner. Another option is to construct a biofilter to which the biogas is discharged, which absorbs and biologically oxidizes the odorous gases. This approach, however, probably does not significantly reduce the methane load on the environment, which is a growing concern in global warming. Methane is among the gases contributing to this change in the global environment.

More recently, floating permeable covers have been suggested as hav-ing potential to reduce the odor escaping from anaerobic lagoons. This process is an engineered approach to accomplishing what dairy farmers had previously observed; that if bedding and other fibrous material in wastewater floats to the surface and forms a dry crust on the top, the odor is effectively reduced. This process prevents the sweeping away of odor-ous gases and there are abundant opportunities for biological oxidation of odorous gases as they permeate the crust. Straw and oil-treated straw have been blown on anaerobic lagoon surfaces in some areas and effec-tively reduce odors. Synthetic materials to accomplish the same result are being tested. Experiments with floating straw and synthetic covers sug-gest that a permeable cover can significantly reduce the emission of am-monia and other odorous gases. It is not clear how much of the effect is related to the biological activity within the cover versus the shielding of the liquid surface from air movement.

Aerated lagoons for odor control. The addition of one or more aerators to anaerobic lagoons has been used as a technique for reducing odor and improving effluent quality in many locations. Often, the aeration capac-ity has been less than would be calculated to maintain aerobic conditions throughout the lagoon contents. For example, some designers have added sufficient oxygen transfer capacity to meet half the daily 5-day BOD load. This is 25% of what is recommended for a totally aerobic sys-tem. One explanation is that the aerators are designed to mix and aerate

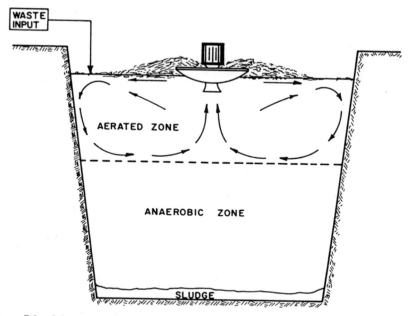

Figure 7.6. Schematic of a mechanically aerated lagoon. The floating mechanical aerator is designed to maintain an aerobic layer at the surface of the lagoon.

only the upper portion of the lagoon. Floating surface aerators typically transfer about 3 pounds of oxygen per horsepower-hour. A second function of the aerators is to provide enhanced mixing. A mechanically aerated lagoon is shown in Figure 7.6. Both the aeration and the mixing functions can be expected to enhance aerobic decomposition of the intermediate breakdown products of anaerobic decomposition, hence reducing odor production. Aeration also can be expected to improve the quality of lagoon effluent for reuse as a flushing medium if effluent is taken from the upper layers where aeration is effective.

If aeration is to be added to an existing anaerobic lagoon to achieve odor control, it would be wise to add sufficient aeration capacity to meet half the daily 5-day BOD loading. The number of individual aerators should be based on the manufacturers recommendations with regard to spacing to achieve effective aeration of the entire surface. If diffused aeration is adopted through a distribution pipe, the piping system should be designed to mix and aerate the entire surface. Thus, a square or round lagoon is easier to aerate than a rectangular one with a large length-to-width ratio. A deep lagoon probably is more successfully aerated than one less than 10 feet in depth because a deeper lagoon maintains an effective anaerobic zone in the lower portion for the breakdown of solids.

There is little biological treatment in lagoons when the temperature is below 40°F, thus aerators are generally shut down in the early winter when the water temperature falls below 40°F. They should be started again in the spring after the ice breaks up and before active anaerobic decomposition begins.

Multiple-celled lagoons. Where a higher quality effluent and less odor release are desired for reuse as flushwater, multiple-celled lagoons can be helpful. The suggested practice is to design the primary lagoon according to the guidelines for a single-celled lagoon but to incorporate the storage function and an additional 25% treatment volume in a second cell. In this way, solids tend to remain in the initial cell where decomposition is enhanced by maintaining a constant water depth. The second cell usually maintains a large algal population, further enhancing the quality of the effluent. Aeration equipment also can be added to the second cell if further improvement in effluent quality is needed.

If multiple cells are to be constructed, they can be placed at different elevations if this arrangement better fits the terrain of the site. Transfer lines should be constructed to take effluent from the primary cell beneath the surface to eliminate floating scum but in the upper 2 feet of the contents to get the best possible quality of water.

Naturally aerated (facultative) lagoons or waste stabilization ponds. Naturally aerated, facultative lagoons, sometimes called waste stabilization ponds, have been widely used in much of the United States for the treatment of domestic sewage. They are particularly popular for smaller communities or in locations where large expanses of land are available and where a conventional mechanical treatment system would be difficult to finance and operate. A cross section of a typical waste stabilization pond is shown schematically in Fig. 7.3.

Lagoons of this design have not achieved widespread application in the treatment of livestock and poultry waste because of the large land area required and the difficulty of maintaining appropriate water levels. In areas where annual evaporation exceeds rainfall rate, lagoons of this design typically lose water due to evaporation faster than water is added by influent. Thus, supplemental water may be needed. In the design of a naturally aerobic lagoon, therefore, performance of an accurate water balance calculation is critically important.

Suggested loading rates for facultative lagoons are much lower than for anaerobic lagoons. Although the loading rate varies with climatic conditions, the suggested daily loading rate in Missouri of 35 pounds of

5-day BOD per acre is typical. Thus, a lagoon of this design would require approximately 1 acre of surface for each 100 finishing pigs. Depths are typically 3 to 6 feet to allow for the necessary surface oxygen transfer and for algae to be effective throughout much of the volume. Other design considerations are similar to those suggested for anaerobic lagoons. For comparison, an anaerobic lagoon at the suggested daily loading rate for Central Missouri of 1.1 cubic feet per pound for 100 hogs averaging 135 pounds would be 149 cubic feet or a surface area of approximately 2,500 square feet or 0.06 acre. This demonstrates the greatly reduced land requirement for anaerobic lagoons.

8

Aerobic Treatment

Aerobic treatment processes depend upon microorganisms that require the presence of dissolved free oxygen to sustain their metabolic processes. Aerobic treatment is common in the municipal wastewater field as trickling filters and a variety of activated sludge options. Each of these treatment systems also has been applied to livestock waste but due to the additional cost involved in maintaining an adequate oxygen supply, most livestock producers have selected anaerobic treatment options. There is a renewed interest in aerobic treatment, however, as livestock operations become larger, environmental constraints demand a greater degree of protection for surface and groundwater resources, and surrounding residents become less tolerant of odors emitted from anaerobic lagoons and other manure storage and treatment devices.

During the past 30 years in which confinement livestock production has grown from an experimental to a standard practice, aerobic storage and treatment devices have been used primarily for their odor control advantages. When storage of manure in pits beneath slotted floors was recognized as causing odorous and potentially toxic gases to enter the animal confinement areas, aeration of these pits was an attractive alternative that reduced both the odor and the presence of ammonia and hydrogen sulfide in the building. A natural outgrowth of this observation was the development of the equipment to operate underfloor oxidation ditches. Underfloor oxidation ditches were developed for swine, cattle, and even poultry confinement houses. As the cost of energy increased between 1970 and 1980, however, the operating cost of oxidation ditches caused them to fall into disfavor. A series of systems evolved that removed the manure from buildings to an outside storage and treatment area, thereby improving the building environment and eliminating the concerns about animal productivity and worker safety.

More recently, as land for manure application has become limiting in many areas and the odors from anaerobic lagoons less acceptable, the benefits of aerobic treatment processes as part of a larger treatment sequence are being rediscovered. Today, aerobic treatment is viewed as a potential supplement to anaerobic digestion for reducing odor and ammonia volatilization and as a component in the process to achieve an effluent suitable for discharge. Another application of aerobic treatment is to reduce the nitrogen concentration by denitrification in anaerobic sequencing biological reactors (ASBR) that have alternating aerobic and anaerobic conditions, thereby reducing the land requirement for effluent disposal. Aerobic treatment also is being used as a means of reducing the odor associated with recycling anaerobic lagoon effluent for flushing manure from buildings. Swine producers in Taiwan, for example, include aerobic treatment as one component in their strategy to produce an effluent from pig farms that meets current discharge requirements.

Aerobic Treatment Process

Aerobic treatment is a process whereby aerobic bacteria use dissolved and suspended organic matter as a source of energy to produce additional bacterial mass and to conduct normal biological functions. The systems available for treating agricultural wastes aerobically involve contacting the liquid wastes with the aerobic bacteria in an environment in which dissolved oxygen is present. Systems for aerobic treatment of animal wastes arose from systems that were already being used in the treatment of municipal and industrial wastes. There are two basic types of aerobic treatment: fixed film processes and suspended growth processes.

Fixed Film Processes

Trickling filters and rotating biological contactors are examples of fixed film processes. In a trickling filter, wastewater with dissolved organic matter is intermittently sprayed over a stone-filled reactor. The stones develop a covering layer of bacteria that feed on the organic matter contained in the wastewater wave that comes along every few minutes. This layer of bacterial cells continues to grow on the surface of the media, becoming thicker with time. The stones are contained in a cylindrical container with air vents along the outside that distribute the air across the bottom of the filter. This air then moves upward through the bed of stones at the same time as the liquid wave is moving downward. The bacterial slime layer thus has the benefit of a supply of both food and oxygen. This process continues until the layer of microorganisms in contact with the stone no longer receives an adequate supply of food and oxygen. At this point, the microorganisms die and the slime layer is sheared off by a subsequent wave of wastewater and is carried out of the filter to a settling basin where the solids settle and are removed for alternate treatment, usually anaerobic. The fixed film process is shown schematically in Figure 8.1.

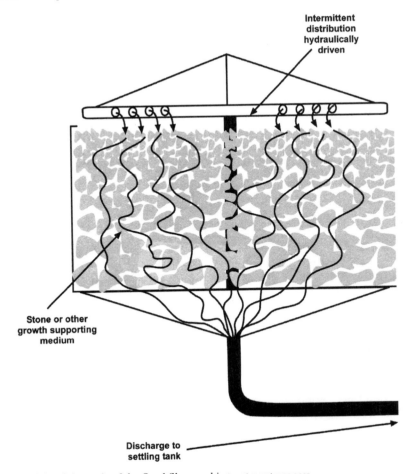

Figure 8.1.   Schematic of the fixed film aerobic treatment process.

   This process is highly stable and in practice converts soluble organic matter, typically measured as biochemical oxygen demand (BOD), into bacterial mass that can be removed in a settling tank. The design of trickling filters for municipal and many industrial wastes is well-established and is based on both an organic loading expressed as mass of BOD per unit volume of filter and a hydraulic loading expressed as volume of liquid per area of the filter. Trickling filters typically remove 75 to 90% of the applied organic matter. The higher removal rates are achieved by recycling a portion of the settled effluent back over the filter.

The rotating biological contactor (RBC) works similarly except the medium is in the form of disks approximately 8 to 12 feet in diameter spaced along a horizontal shaft that revolves at a slow speed, alternately contacting the surface with wastewater and air. The bacteria that comprise the active mass behave similarly to those on the trickling filter stones. RBCs have been used experimentally in the treatment of livestock waste and were found to operate satisfactorily but were more expensive to purchase than other alternatives. In addition, the high BOD concentrations required considerable recycling to not overtax the oxygen transfer capability of the system. Trickling filters and RBCs are always operated in conjunction with a settling tank that follows to remove the synthesized bacterial mass.

Suspended Growth Processes

Suspended growth systems of aerobic treatment are typically called activated sludge systems because the bacteria are not attached to a fixed surface but are instead suspended in the wastewater being treated. After a suitable contact period, they are separated by settling and a portion is recycled back to the treatment tank. The treatment tank is aerated to supply the oxygen required by the bacteria and to keep the cells in suspension so they come into contact with the food supply. The name "activated sludge" seems to have arisen from the practice of aerating the recycled sludge after it was recovered from the wastewater as a means of preparing, or activating, it to more effectively metabolize organic matter and convert it to additional biomass. For the trickling filter, a settling tank typically follows an activated sludge process for removal of synthesized biomass. The suspended growth aerobic process is presented schematically in Figure 8.2.

Numerous modifications of the activated sludge process have arisen in response to the need to achieve different levels of treatment and to meet the needs of various municipal and industrial wastewater treatments. The common characteristic of all these processes is the use of a suspended biological mass in contact with the wastewater being treated. Typically, this contact occurs in an aerated tank to facilitate the metabolism of organics. In many of the modifications, the separated sludge is subjected to aeration before being reintroduced to the treatment tank to restore its ability to immediately absorb organics. Thus, one of the modifications called contact stabilization stresses this aspect of the process and includes a very short contact between sludge and the wastewater but more intensive aeration of the recycled sludge.

Conventional activated sludge treatment is essentially a plug flow process in which recycled sludge contacts incoming wastewater and the

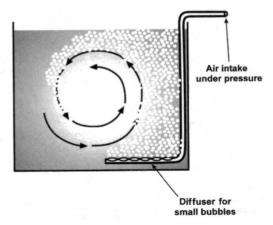

Air intake
under pressure

Diffuser for
small bubbles

Figure 8.2.   Schematic of the suspended growth aerobic treatment process.

combination remains in contact for 4 to 6 hours of aeration. From the aeration basin, the mixture flows to a settling basin. The sludge is recycled and the supernatant is discharged as effluent that may or may not require additional treatment. The modification that received the greatest application for treatment of livestock waste was the extended aeration process in which incoming wastewater and recycled sludge were held in an aerated tank for 24 to 48 hours then settled. The effluent was either stored for eventual land application or recycled as flushwater. Oxidation ditches and aerated lagoons represent adaptations of the extended aeration process. These processes have received widespread application in smaller municipal wastewater treatment because of their high stability and adaptability to commercial manufacture as a pre-engineered package. None of the extended aeration options is being extensively used at this time due to the availability of low-cost alternatives that provide equivalent treatment.

Aeration Process

At the heart of any aerobic wastewater treatment process applied to livestock and poultry waste is aeration. Because oxygen is only slightly soluble in water, 7 to 10 milligrams per liter, and the organic content of most liquid animal wastes is many times higher, an outside oxygen source is necessary. Two types of aeration are possible: diffused aeration in which air is blown in small bubbles into the wastewater; and mechanical aeration in which an aerator is installed that beats, pumps, throws, or otherwise creates a large surface area of wastewater in contact with the air so that oxygen can be absorbed to replace that consumed by the bacteria as they metabolize the soluble organic wastes. Whatever the mechanism for introducing oxygen into the liquid, it follows the relationship

$$dC/dt = KLa \ (Cs - C)$$

where dC/dt = rate of change of the dissolved oxygen concentration with time, mass/volume–time; KLa = overall gas transfer coefficient, time^{-1}; Cs = oxygen saturation concentration for the existing atmospheric composition, temperature, and pressure (mass/volume); and C = actual dissolved oxygen concentration existing in the liquid (mass/volume).

There are several important aspects of this equation. First, it reminds us that it is impossible to aerate a liquid to an oxygen concentration beyond saturation. Second, it indicates that the greater the dissolved oxygen concentration, the more difficult it is to add oxygen. The equation also reminds us that both the mass transfer coefficient and the saturation concentration are dependent upon temperature and the presence of other materials in the water that is being aerated. Furthermore, this equation is used in the evaluation of aeration equipment to determine oxygen transfer capacity.

Aeration equipment is generally evaluated by placing the equipment in a tank of water from which most of the oxygen has been removed by chemical means. Sulfite reacts quickly and irreversibly with oxygen in the presence of a cobalt or copper catalyst. The aerator is then operated and the dissolved oxygen concentration measured as a function of time. These data are plotted as the logarithm of Cs − C versus time. The data plotted in this manner lie along a straight line. The slope of the line is KLa. By knowing the volume of the tank and the water temperature, it is then possible to calculate the oxygen transfer capability of the equipment.

One of the differences between the aerobic bacterial process and the anaerobic process is that the aerobic process yields a greater quantity of energy to the bacteria for each unit of organic matter processed. Thus, the aerobic process results in a greater production of bacterial cells that either accumulate, settle to the bottom, or must be removed for alternate use or disposal. In aerobic treatment, approximately 50% of the carbon goes to biomass, whereas in anaerobic treatment approximately 10% goes to biomass, with the remainder of the carbon, in each case, being converted to carbon dioxide or methane. These bacterial cells have been evaluated as a potential feed material and they are certainly appropriate

Floating Surface Aerator

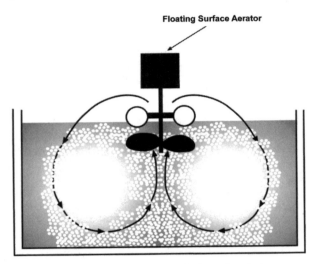

Figure 8.3.   Schematic representation of mechanical aeration device.

for this use, but because they are mixed with such a large quantity of water, they are less attractive for refeeding than would otherwise be the case.

In addition to supplying the necessary oxygen to maintain the aerobic process, aeration equipment provides mixing of the material. This mixing is important in the treatment of livestock and poultry manure if the goal is complete aerobic treatment. If the goal is surface aeration only, to produce a lagoon with an aerobic surface free of odor and ammonia volatilization, yet preserve the benefits of anaerobic digestion of the bottom sludge, some kind of shield is generally provided beneath the aerator to limit the depth of mixing.

Aeration also tends to strip volatile material from the liquid being aerated. If the material being aerated enters the aeration chamber with a significant concentration of dissolved ammonia and the pH is above 7.0, a major portion of that ammonia is stripped from the liquid and dispersed into the air.

Two major types of mechanical aeration systems are available: diffuser and mechanical. In the diffuser system, air is forced under pressure to escape to the liquid phase through a porous surface or a specially designed fitting to create multitudinous small bubbles (Fig. 8.2). Mechanical aeration systems create turbulence at the surface so that air contact with the liquid phase is enhanced (Fig 8.3). In either system, the goal is to create a sufficiently large water–air contact surface to facilitate the required oxygen transfer rate.

Application of Aerobic Treatment Processes

There are currently two major applications of the aerobic biological treatment process relative to livestock waste management. One is in the storage of manure prior to land application. The other is in a treatment scheme in which the effluent is being prepared for reuse within the livestock operation, or for land disposal. Aerobic systems are particularly useful where there is insufficient land available for spreading based on nutrient conservation. In some parts of the world, manure-laden waters are being treated for discharge into receiving streams. Although this latter option does not have current approval in the United States, there are portions of the country in which the amount of manure being produced exceeds the capacity of nearby cropland to effectively use the nutrients. There are also areas in which the problems inherent in existing manure treatment systems, such as odors and ammonia volatilization from anaerobic lagoons and spray fields, and disastrous lagoon dike failures, are prompting a rethinking of the traditional approaches to manure management. Under these conditions, an alternate final disposition is sought. Stream discharge is one alternative. Nutrient removal to decrease the amount of land required is another alternative. Under either of these conditions, aerobic treatment is likely to be a component of many of the systems that evolve.

Aerated Manure Storages

Aerated manure storages are generally constructed where it is important to achieve a high degree of odor control. Aerated manure storages can be designed to conserve a major portion of the nitrogen by conversion of ammonia to nitrate or by periodic interruption of the aeration, nitrogen can be removed. Whether nitrogen is conserved or vented to the air is a function of whether the nitrogen has an economic value for use as a plant nutrient in the specific location. Aerobic storage can be in an aerated reservoir beneath the building or it can be in a tank or basin outside of the building.

Oxidation ditch design. An oxidation ditch is generally built as an underfloor manure storage basin that is aerated for odor control and to maintain a quality environment within the building for the housed animals. The unique aspect of an oxidation ditch is a configuration that generally involves a dividing wall down the center of the tank, creating a racetrack-shaped container in which the manure is stored. Aeration and mixing are provided by a horizontally shafted aerator mounted so the shaft and motor are above the liquid level. The rotor is constructed similar to a street sweeper and extends into the liquid, providing both aeration and movement to the liquid.

The design of an oxidation ditch should proceed in much the same way as other manure storages by deciding the duration of the storage, the number of animals to be served, and the daily input volume per animal. In addition, it is important to remember that the ditch must be filled with water up to the minimum operating level before manure is introduced for the oxidation ditch to function. This initial charge of water is generally 12 inches. Depending upon the ability of a specific rotor to provide velocity to the stored waste, maximum depths need to be specified. Two feet is a typical maximum depth. Clearly, some means of adjusting the height of the rotor is required as the liquid level increases to maintain optimal submergence of the rotor blades. A schematic view of an under-building oxidation ditch is shown in Figure 8.4. Some typical design parameters for an under-slotted floor oxidation ditch are presented in Table 8.1.

A proper interpretation of Table 8.1 is that an oxidation ditch beneath the floor of a livestock confinement facility is an initial design decision that needs to be incorporated from the beginning. It is seldom possible to retrofit an existing building to incorporate an oxidation ditch. Table 8.1 also indicates that an individual oxidation ditch serves a relatively small number of animals. For example, assuming an average animal size of 120 pounds and an oxidation ditch width of 7 feet, each 6-foot rotor would accommodate approximately 100 animals. Up to four rotors have been installed equally spaced around an oxidation ditch.

Oxidation ditches have been constructed in applications other than beneath slotted-floor confinement buildings; however, other methods of aeration and other configurations of storage are more popular for these applications. One option is worthy of mention. As originally used in Europe, the oxidation ditch was operated as a continuously aerated tank for approximately 20 hours then the rotor was stopped and the suspended biomass allowed to settle. After an appropriate settling time, effluent was discharged. Once the oxidation ditch level had returned to the target level, the discharge valve was closed and the aerator activated. This operating scheme has been used in the treatment of livestock waste and has particular advantages. First, it achieves both aeration and settling in a single tank. Second, it has the potential of achieving nitrogen removal by oxidizing the ammonia to nitrate during the aeration phase and allowing denitrification to occur while the aeration is interrupted. Although this concept of using a single basin for both aeration and settling is not

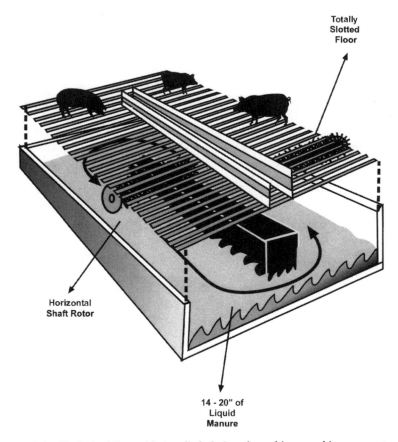

Figure 8.4. Under-building oxidation ditch designed to achieve aerobic manure storage.

Table 8.1. Typical design parameters for the design of an under-building oxidation ditch for the storage of manure from a confinement pig raising building

Parameter	Value
Operating volume of the ditch; this volume should be initially filled with water	0.065 cu ft/lb pig served
Initial water depth	12 in.
Oxygen transfer capability	0.0042 lb/day/lb pig served
Length of rotor required	0.0005 ft/lb pig served
Anticipated power consumption	0.0022 kW-h/lb of pig served based on an oxygen transfer effectiveness of 1.9 lb. oxygen transferred per kW-h

exclusively suited to the oxidation ditch configuration, it is appropriate to consider in this context.

Aerobic manure storage basin and tank design. As with an oxidation ditch, an outdoor basin or tank can be used for aerobic storage and treatment of manure. There are several options. One is to discharge the total manure from a building either by flushing or by using one of the fill and dump systems to flow into an aerated basin designed to hold the manure flow for a selected time period. An aeration capacity would be selected to provide approximately twice the daily BOD contribution from the animals. This particular system would provide a low-odor means of manure storage that could be operated to minimize nitrogen loss or if desired, it could be operated to deliberately achieve nitrogen removal based on nitrification and denitrification.

A second possibility would be to divert the raw waste flow over or through a solid–liquid separator and then to store the liquid fraction in an aerated basin or tank. By removing the solids prior to aeration, less aeration capacity (15 to 25%) would be required and the accumulation of solids in the aeration basin would be greatly reduced. This strategy is particularly attractive where there is a market for stored or composted manure solids. Dairy operators, particularly those located near metropolitan areas, have discovered a market for composted manure solids among gardeners and ornamental nursery operators. The aerated liquid from a storage tank of this design would be suitable for crop or pastureland application and for reuse as a flushing liquid. It would not be suitable for stream discharge but it might be appropriately discharged to an engineered wetland for nitrogen removal and polishing treatment.

Aerobic Treatment

In contrast to the above-mentioned systems that were intended to facilitate use of manure nutrients on cropland to recover their economic value, there are a growing number of situations in which alternate strategies are being sought. Many of these alternatives involve one or more aerobic treatment components. For example, in Taiwan and Korea, there is essentially no land available to the pig and poultry producers for manure application because neighboring farms are small and it is difficult to maintain cooperative arrangements to use animal wastes as a crop fertilizer. Thus, waste handling systems are being designed to use alternative disposal. These systems are treating wastes for discharge to surface streams. A second group of operators likely to incorporate aerobic treatment into their overall strategies are those who find that the current storage and land application technologies are resulting in unacceptable odors or those that do not having adequate land available for manure application. Aerobic treatment will probably be part of many of these systems in the United States as they are currently in Asia and eastern Europe.

Aeration tanks or ponds. The simplest and probably most popular means of aerobic treatment is an aerated tank or earthen basin to provide additional treatment and odor reduction to the effluent from an anaerobic digester or lagoon. Where effluent is being used as flushwater and odors in the buildings and in the surrounding community necessitate further treatment, simple aeration is an alternative. The disposal of lagoon effluent with sprinkler irrigation equipment is a low-cost means of distributing the effluent from anaerobic lagoons, but ammonia volatilization and odor release make use of this equipment unacceptable in some locations without a terminal aerobic treatment.

The design of an aeration tank or aerated earthen basin to serve a particular operation depends to some extent on the system into which it fits and on the intention of the manager. If the goal is to achieve flushwater that has been stripped of the easily volatilized odorous compounds and made aerobic, the required storage time in the aerator is relatively short. One option is to construct an aeration tank with sufficient volume to provide a 1-day supply of flushwater. The aerator should have sufficient capacity to provide twice the maximum daily BOD load calculated by having the BOD determined during late spring when the lagoon is likely to be least effective in BOD reduction. If the aerator proves to be capable of maintaining a dissolved oxygen (DO) concentration in excess of 2 milligrams per liter during other seasons it can be run intermittently by installation of a timer. There is little benefit in maintaining a DO concentration in excess of 2 milligrams per liter. If it is not possible to base the aerator selection on a measured BOD concentration that represents the maximum value, an estimate can be used. This estimate would likely be based on 15% of the concentration calculated by dividing the BOD production (pounds per animal) by the amount of flushwater used per animal expressed in pounds [(gallons per day) (8.34 pounds per gallon)]. For example, if the representative animal is a 150-pound finishing pig that produces 0.30 pounds of BOD per day and the operation uses 10 gallons of flushwater per animal per day, the concentration leaving the building would be approximately (0.30 pounds BOD/pig-day/(10 gallons/pig-day $\times$ 8.34 $\times$ 10^{-6} million pounds per gallon). Lagoon effluent BOD concentration would be estimated at 550 milligrams per liter for purposes of aerator selection. Continuing the example, based on the 10 gallons of flushwater per pig per day would indicate an aerator capable of providing 2(550 milligrams per liter)(10 gallons per pig-day)(8.34 pounds/gallon)(10^{-6} pounds/million pounds) = 0.10 pounds of oxygen transfer capacity per pig per day. This amount of aeration is only 15% of that required if totally aerobic treatment had been selected as the option.

There are other options for the use of aeration tanks, depending upon the overall treatment strategy. For example, if aeration is to be used as part of a nitrogen removal strategy, it should probably follow an anaerobic digestion phase. Under these conditions, the goal is to quickly establish aerobic conditions and a population of aerobic nitrifying organisms to convert available ammonia to nitrate that can subsequently be denitrified with nitrogen gas vented to the atmosphere. In this case, laboratory analyses are necessary to size the aeration equipment being selected.

Other aerobic treatment devices. Although aerobic treatment of livestock wastes has largely focused on the simple aerated tank or basin, other alternatives exist and have been used. Trickling filters, common in the treatment of municipal wastes, can be used. Trickling filters also are used in the treatment of many organic industrial wastes. In practice, a trickling filter appears as a cylindrical tank containing 7 to 10 feet of medium over which the liquid waste is discharged. The medium, frequently granite stones, nominally 2 to 5 inches in their largest dimension, is selected to provide extensive open area so that air can move freely through the full depth to maintain aerobic conditions. The wastewater is applied to the trickling filter after the solids have been removed, usually in a settling tank. Liquid application is usually accomplished with a rotating distributor system with two or four arms. As the distributor rotates, each section of the medium is dosed with liquid waste. A mixture of effluent and sloughed biomass flows from the bottom of the filter to a settling tank. Figure 8.5 is a representation of a typical trickling filter process. It is typical that a portion of the settling tank overflow is recycled to the trickling filter to maintain a constant hydraulic loading rate. High recycle rates also increase the organic removal rates. Typical design characteristics of trickling filters are shown in Table 8.2.

Trickling filters are particularly common in the food processing industry and are suitable for use as a pretreatment step if waste strength needs to be reduced prior to discharge into a municipal wastewater system. The use of trickling filters to treat settled livestock wastes has been less popular due to the high BOD concentrations. As the industry moves toward greater use of anaerobic pretreatment technologies, lagoons and digesters, trickling filters as secondary treatment become more likely. Trickling filters can be operated to convert a high fraction of the dissolved ammonia into nitrate.

Figure 8.5.   Typical trickling filter process used for aerobic treatment of organic waste materials.

Table 8.2.   Typical trickling filter design characteristics

Characteristic	Value
Depth of media	6- to 10-ft filters up to 25 ft have been used
Nature of the media	Typically weather-resistant stone; however, plastic media are available that provide a high surface-to-volume ratio; redwood slats also have been used
Hydraulic loading rate	10–40 million gal/acre/day; both higher and lower rates may be selected
Organic loading rate	30–100 lb of BOD/1,000 cu ft/day.
Expected BOD removal	65–90% depending upon loading and recycle rates; lower loading rates along with higher recycle rates contribute to higher removal efficiencies

There are other configurations of aerobic treatment being used in the management of livestock and poultry waste. Short-term aeration of a few hours to a few days is used as an odor control and ammonia stripping process. In relatively mild climates where liquid manure can be applied to crop or pastureland throughout the year, it is possible to use an aerated tank or basin sufficient for 1 to 3 days of storage. Solid–liquid separation is common before such an aeration tank. Combining aerobic and anaerobic treatment to achieve both organic matter and nitrogen reductions is possible. One of the devices available is the ASBR. This system is discussed in greater detail in Chapter 7 but the basic concept is to alternately aerate the liquid waste and to allow it to stand quiescent without aeration.

Under appropriate conditions, the BOD is reduced and the nitrogen oxidized to nitrate during the aerobic phase, then it is reduced to nitrogen gas during the anaerobic phase. Properly designed and operated, the ASBR is a useful tool for preparing the liquid manure for reuse in a flushing system or as part of a larger system to achieve an effluent that can be spread on a more restrictive land base.

Aeration Equipment Available

There are three basic types of aeration equipment available for use in aerobic treatment of livestock waste. The first type is the diffused air system in which air under pressure is allowed to enter the liquid through a porous plate or tube or through perforated pipe. The diffusers, as they are called, are placed at a liquid depth of up to 10 feet and the air bubbles float upward through the liquid with a portion of the oxygen dissolved in the water. The second type is the sparged air turbine system. In this system, compressed air is allowed to escape from a diffuser or series of perforations directly beneath a mixer blade located several feet beneath the liquid surface and spinning at a high speed. The third system is the surface entrainment aerator that mechanically creates turbulence at the surface with the rapid movement of partially submerged blades. Surface entrainment aerators may be either vertical shafts driven such that the water is thrown radially away from the shafts in all directions or there are horizontal shaft aerators, typically called rotors, that throw the water in a single direction (the direction the blades move when submerged). Oxygen transfer capabilities of mechanical aerators are dependent upon design as well as construction detail. Manufacturers can make this information available. Designers can anticipate oxygen transfer rates ranging from 2 to 4 pounds of oxygen transfer per kilowatt-hour.

Diffused aeration in which bubbles are released at the bottom of the tank and float to the surface creates a water velocity and a mixing pattern that resembles a rolling of the tank contents. The rate of oxygen transfer through these systems is a function of the turbulence created and of the air–water interfacial area. Thus, many of the early diffusers had very small holes. Even porous ceramic pipes were used for aeration. The small holes and porous materials have the disadvantage of being prone to plugging due to the formation of chemical precipitates or more typically to the development of a tenacious biological film. In either case, considerable effort is required to maintain airflow without major increases in the pressure required. Current diffuser systems tend to have larger openings to reduce plugging. The oxygen transfer efficiency of diffused aeration systems is approximately 5%, ranging from 3 to 8%. Diffused aeration equipment is currently more popular in municipal wastewater than in livestock waste treatment because of the availability of maintenance personnel in municipal plants. When compressed air is used to provide oxygen for transfer to a liquid system, it is customary to refer to oxygen transfer capacity as the fraction of oxygen absorbed by the liquid, divided by the total amount of oxygen pumped or compressed. Both amounts are commonly stated in pounds per hour.

The submerged turbine systems have separate mechanisms for introducing air and for mixing functions. In this system, air bubbles are introduced beneath the spinning turbine through either a single sparger or through a series of holes or release mechanisms arranged in a circle. Thus, as the bubbles rise they enter the area of the spinning turbine where they are sheared into smaller bubbles, creating a much larger and a freshly formed air–water interface. It is possible with this system to balance the need for oxygen and for mixing because the spinning turbine provides mixing independently of air supply. Typical efficiencies for a submerged turbine aerator may be in the range of 15 to 20% but higher efficiencies are possible. Submerged turbines transfer more oxygen to a specific volume of wastewater than either of the other two options and in addition, have fewer potential problems due to ice formation on the blades than the surface aerators. The disadvantage of these systems is that they require two mechanisms to maintain and the difficulties inherent in controlling plugging problems associated with submerged air inlets.

Surface entrainment aerators, commonly called surface aerators, are the most popular style for the treatment of agricultural waste. Surface aerators achieve the necessary air–water contact by means of an impeller or turbine located at the liquid surface. Most surface aerators used in the treatment of livestock and poultry waste are sold with a floatation mechanism as part of the basic device so that the aerator rides on the surface of the tank or pond. Aerators also are available that are attached to fixed supports in an aeration basin. These units are suitable only in systems in which the water level is held constant because aerator performance is very sensitive to the depth to which the blades are submerged. Most surface aerators have vertical drive shafts directly connected to the electric motor providing the power. Thus, they are simple mechanisms both to manufacture and to maintain. Surface aerators tend to be highly effective in mixing in the immediate vicinity of the unit but may not effectively mix the entire basin or pond. In addition to specifying the oxygen transfer capability of an aerator, the designer needs to consider the effective mixing radius of the aerator. It may be necessary to specify several smaller aerators rather than a single larger one to provide the required degree of mixing.

The efficiency of surface aerators is typically measured as the pounds of oxygen transferred to dissolved oxygen-free water at 20°C per horsepower hour. The efficiency typically ranges from 2 to 4 pounds of oxygen per horsepower hour. Three is a standard design value. The value depends upon the design of the aerator and the speed at which it is driven. It is also important to note that in ponds fitted with surface aeration, almost all of the aeration comes from mechanical aerators rather than from a combination of natural surface aeration and mechanical aeration.

There are also horizontal shaft aerators, sometimes called brush aerators because of their resemblance to the brush on the front of a street sweeper. The horizontal aerators operate by spinning with the blades of the brush partially submerged so they seem to throw the water in a single direction perpendicular to the shaft. In this way, they create a velocity as well as introducing oxygen into the water. The brush-type aerator is used almost exclusively in oxidation ditches where maintaining an adequate velocity to keep particles in suspension is critical. Brush aerators are typically sized anticipating approximately 25 pounds of oxygen transfer per day per foot of aerator or expressed differently as 2 pounds of oxygen transferred per kilowatt-hour.

9

Composting

Composting is a natural aerobic process for the stabilization of a variety of organic matter ranging from forest litter to manure in stalls of horses and cattle. Composting is one of the major recycling processes by which materials return to the soil in the form of nutrients available for future use. Recently, engineered systems for the conversion of manure from livestock and poultry into compost have become popular. Some of this popularity has been based on the concept of converting a material, manure, from a financial liability into a marketable product. Examples abound in which compost is being produced from manure and other waste materials at a profit, but other examples can be cited in which compost, although easily produced, has not been accepted in the market with sufficient enthusiasm to justify the costs involved. Thus, in this chapter composting is presented as a treatment process that may be incorporated into a manure management plan. Compost as a product has certain advantages over fresh manure or chemical fertilizer for application to cropland, gardens, and lawns.

Benefits of Composting

The composting process changes the physical, biological, and chemical characteristics of the material that is composted. When animal manure is composted, the material has the readily available organic matter stabilized to the extent that it is no longer readily decomposable, hence it is no longer subject to further anaerobic decomposition with the associated odor release. Well-composted animal manure has the odor of humus and is considerably more acceptable for land application in locations where fresh manure would be objectionable. In the stabilization process, the volume is reduced. Actual reduction probably ranges from 25 to 50%, depending upon the initial material. Because of the heat produced during composting, well-controlled composting results in the death of common pathogens and weed seeds. Pathogens and weed seeds are of greater concern with human waste products and sewage sludge, however, a process that can ensure freedom from pathogens and weed seeds is of obvious merit. The physical characteristic of compost is that of

187

rich soil rather than that of the initial material. Some ingredients are more resistant than others to composting. For example, some feathers and feather parts may survive the composting of poultry litter. If dead animals are included in the compost, some teeth, bones, and bone fragments also may be identifiable.

The extent to which nitrogen is conserved in the composting process depends largely on the carbon-to-nitrogen ratio of the feed material. If the feed has a carbon-to-nitrogen ratio of 30:1 or above, little nitrogen loss is anticipated. If the ratio is less than 30:1, as is the case if manure is the principal ingredient, ammonia release tends to raise the ratio. If the feed ratio is 20:1, nitrogen losses may be as high as 40%. Carbon-to-nitrogen ratios above 30:1 generally result in nearly complete nitrogen conservation but may require a longer time to reach completion due to the nitrogen being a limiting nutrient.

Composting is most frequently adopted by livestock producers who anticipate a market for the finished material. This market may be nearby garden and nursery supply outlets, landscaping services, or contractors needing to establish lawns or landscaping in conjunction with completed construction projects. Cities frequently use compost in the establishment and maintenance of parks and other recreational areas.

The advantages of compost over fresh manure are the reduced odor; reduced fly attraction; reduced pathogen and weed seed concentration; and according to many horticultural studies, a better plant response due to the addition of a stabilized organic material that builds soil tilth. Clearly, composted material is more desirable to handle for many home and recreational gardeners. The disadvantages of compost are the additional processing costs and the need to remove and manage a larger amount of solids in the waste management system.

Thus, like other manure treatment alternatives, composting is not compatible with all livestock operations. The design of manure handling facilities requires careful consideration of various alternatives.

Composting Process

Because composting is such a widely observed natural process, occurring throughout the tropical and temperate climates, it should not be surprising that composting can be successfully practiced under a variety of environments. Whenever biologically available organic matter is present, a supply of water, oxygen, and temperatures between zero and 70°C, a population of naturally occurring microorganisms develops that can glean energy from the decomposition process and thus proliferate.

Most composting operations, particularly those selling a finished product for a soil amendment practice, are a two-stage process. The first or active composting stage generally requires 10 to 30 days to complete to the point the material does not heat after turning. At this point, the compost is typically moved to a storage area where it is allowed to age, ripen, and mature. This aging is thought to ensure a more consistent product and allows additional cellulose decomposition. It also is an effective way to deal with the seasonal nature of the commercial compost market. If compost is to be applied to agricultural land, the aging process is probably less important and can be eliminated.

Organisms Involved in Composting

Research over the past 50 years has documented that the composting process whether based upon garbage, night soil, animal manure, or urban vegetation debris involves a large number of different types of bacteria, fungi, mold, and other living organisms. Furthermore, these organisms each tend to provide specific functions for which they are uniquely suited and that there is no single organism or group of organisms capable of conducting the process as effectively as the mixed population that naturally develops in the composting environment.

The organisms necessary for composting are generally available in the materials being composted. The needed organisms tend to thrive under the environmental conditions that exist at the time they are needed. Some of the many species of aerobic and facultative bacterial species multiply rapidly in the early stages of composting but dwindle as the environment changes and other organisms are able to thrive. Temperature and changes in the available food supply seem to exert the greatest influence in determining the species comprising the population at a particular time. The succession of populations reflects constantly changing environments because the temperature and substrate are in a state of constant flux. The substrate changes because of the continual breakdown of complex food materials into simpler compounds. Except for brief interruptions during turning, the temperature increases steadily in response to the energy liberated during metabolism until equilibrium with heat losses is established or the material becomes sufficiently stabilized to be food-limiting.

In the typical manure composting process with livestock and poultry manure mixed with a bulking material such as straw, the facultative and obligate aerobic bacteria, actinomycetes, and fungi are the most active organisms. Mesophilic bacteria initially predominate but are replaced with a thermophilic counterpart as the mass increases in temperature. Thermophilic fungi generally appear after 5 to 10 days and actinomycetes even later when the material has passed its peak temperature. During most of the process, actinomycetes and fungi are found in an outer layer in which the temperature is more moderate than in the center of the pile or trench. The presence of actinomycetes and molds can frequently be seen in the outer layer of a pile as evidenced by a white-to-gray color in the outer 6-inch layer.

It appears that thermophilic bacteria dominate the decomposition process in the center of a pile where temperatures become inhibitory to actinomycetes and fungi. The actinomycetes and fungi have a particularly important role in the decomposition of cellulose, lignin, and other more resistant materials. As a result, the transformation of cellulosic material, particularly paper, may be delayed until the latter stages of composting when the temperature has moderated sufficiently to allow their development.

Throughout the history of engineered composting processes, there has been an interest in the development and marketing of inocula and enzymes to aid the composting process. If the initial waste materials to be composted were sterile, there would be a scientific basis to support the addition of microorganisms. Because that is clearly not the case and because composting processes throughout the world operate successfully without microbial or enzyme additions, inocula and other additives do not seem to be essential in the composting of waste materials, including animal manure.

Temperature

Temperature is an important factor in aerobic composting. Aerobic fermentation releases a considerable amount of heat. The composting material, generally in a highly porous condition, is a relatively good insulator; hence, a sufficiently large composting mass retains sufficient heat from exothermic reactions and elevated temperatures result. High temperatures are essential for the destruction of pathogenic organisms and weed seeds. Decomposition also proceeds more rapidly in a thermophilic temperature range than at lower temperatures. Most operators agree that the optimum temperature is around 60°C (140°F). Above 70°C, there are few thermophilic organisms that can effectively stabilize organic matter.

Figure 9.1 shows a typical temperature curve for the interior of a pile or windrow of composting manure solids. Usually, a temperature of 45–50°C is obtained in the first 24 hours of composting and 60°C is reached within 3 to 5 days. The final decline in temperature is more gradual than the initial rise and indicates the material has become stabilized. A drop in the temperature before the material has become stabilized suggests anaerobic conditions have developed and that aeration is needed. The temperature of the pile can be expected to vary with the

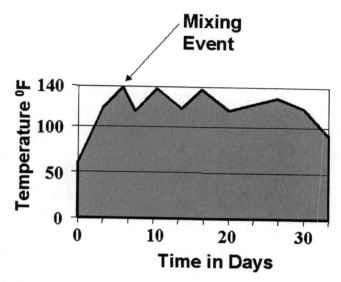

Figure 9.1. Temperature profile in a compost pile or windrow. The equilibrium temperature achieved in the pile or windrow depends upon the size, moisture content, degree of aeration, and ambient temperature.

size of the pile, ambient temperature, moisture content, the degree of aeration, and the nature of the initial feed material. Of all these variables, aerobic conditions are the most important in achieving proper temperatures.

The size of a composting pile or windrow can be adjusted during various seasons to control the operating temperature. Deeper piles are appropriate in cold weather and more shallow piles in hot weather. Experience suggests that turning the pile, which is necessary for uniformly complete composting, is not effective in reducing the temperature of the pile. The temperature decreases during the turning process but is restored within a few hours. Similarly, adding water to the pile is not an effective way to reduce temperature unless sufficient water is added to cause water logging, which is undesirable.

There are several ways to monitor temperatures in a composting mass. The most common way is to insert a long-stem metal thermometer into various parts of the pile (Figure 9.2). An alternative is to insert a metal rod at least 2 feet into the pile. After 10 to 15 minutes, the rod should be too hot to hold comfortably for an extended period. Similarly, the condition of the pile can be evaluated by digging 2 to 3 feet into the pile. The material should be too hot for you to leave your hand inserted into the pile for an extended period. If a compost pile fails to

Figure 9.2.   Interior temperature in a compost pile or windrow can be monitored by inserting a long-stem thermometer into the center and maintaining a log of the readings.

reach a high temperature in 3 to 6 days, it is either too small to retain heat, too dry or too wet, or insufficient organic material or nutrients are available to support an active aerobic decomposition.

Moisture Content

Moisture content is critically important in the composting process for two reasons. First, sufficient moisture must be available for the microorganisms to grow. Without water, bacteria and fungi cannot maintain their activity. Second, bacteria and fungi need oxygen to function. Excessive moisture displaces the air that would otherwise be in a pile, causing the microorganisms to live in an anaerobic environment.

For composting most materials, the moisture content should be between 40 and 60%. The upper limit is somewhat dependent upon the material being composted. If the material being composted contains straw, for example, it may be possible to operate successfully well above 60% moisture because straw retains its strength and turgor at higher moisture contents and stills allow air to move freely through the pile. Waste paper, in contrast, becomes soggy at 60% moisture and packs with sufficient density to exclude air. Thus, a compost pile containing waste newsprint would need to have low moisture content.

When manure is composted, a bulking agent such as straw, wood shavings, or sawdust is typically added to adjust the moisture content and to introduce some structure to the composting mass that traps air and allows air movement within the pile between turnings. If the moisture content is too high, there are several alternate solutions. One is to add a dry, fibrous material to lower the overall moisture content; a second is to turn the pile more frequently during the early stages of composting to incorporate air; and a third alternative is to add mechanical aeration. Once active composting is underway, moisture is lost from the pile and it may be necessary to add water while turning the pile.

The moisture content of compost can easily be determined in a laboratory by drying a previously weighed sample and then reweighing the dry material left after the water is driven off. This method of moisture determination is seldom necessary because the operator of a composting operation soon learns to judge by sight or texture when the moisture content is within an acceptable range.

Aeration and Turning

Aeration is necessary for thermophilic aerobic digestion to produce quality compost and to avoid nuisance conditions during the composting process. Aeration also can be used to overcome an initial moisture content that is too high.

The frequency of aeration or turning and the amount of aeration or the total number of turns required are determined by the moisture content and the material being processed. If the moisture content is high, there is little space available for air within the pile and little opportunity for air to move within the interstitial spaces. Materials with a high carbon-to-nitrogen ratio may need to be aerated less frequently than those with a lower ratio and a more active composting process. Each operator probably determines the optimal turning frequency for his or her particular operation, but an appropriate planning frequency would be every 2 to 3 days, assuming the moisture content is in the 40 to 60% range. If the moisture content is greater than 60%, daily turning probably is necessary until the moisture content decreases. If the moisture content is less than 40%, it is advisable to add water.

If a compost pile becomes anaerobic, indicated by a drop in temperature during the first 7 to 10 days or by a septic odor when the pile is disturbed, turning is required. No matter how anaerobic a compost pile becomes, it can recover with daily turning and aeration.

There are a variety of mechanical devices that provide highly mechanized and more automated composting. Figure 9.3 shows a typically labor intensive process being used to turn and mix composting material.

Figure 9.3.   Turning is an essential part of the composting process to subject the entire mass to the benefits of high-temperature biological processing. Depending upon the volume of compost being processed, there are several alternatives in the choice of equipment.

In more highly automated devices, the compost is continuously mixed and aerated to achieve rapid composting. These devices reduce the labor input into the process and reduce the land area required.

Climatic Conditions

Composting can be conducted under a wide range of climatic conditions but temperature, rainfall, and wind influence the process. Compost is generally a good insulator so the interior temperature of the compost may not be highly impacted. Pile depth may be increased slightly to better conserve heat. Strong winds also can contribute to excessive heat loss. Windbreaks may be necessary in some locations. Smaller particle sizes also provide greater protection against wind penetration.

Rain usually does not interfere with composting operations if the tops are shaped to promote runoff rather than penetration. It is also important that the composting area be well drained to allow runoff to leave the composting area rather than be absorbed by the active piles. Runoff from manure composting areas should be treated as other manure-laden waters and should be collected and either land-applied or treated prior to being released to the environment. In general, turning should not be done during intense rainfall due to the potential for the compost to become waterlogged.

Severely cold weather can be expected to slow the composting process. The most obvious solution is to have additional space available and to plan to turn the material less frequently. The other alternative that has been adopted in several composting operations is to provide a roof and windbreak for the active composting operations. The maturing or aging of compost can generally be done outside even in areas of severe winter weather.

Fly Control

Fly control is one of the more important aspects of compost production for the process to be acceptable. Manure is a particularly attractive medium for fly propagation, hence their control is a necessary consideration in the design and operation of a composting facility. Fly propagation is not an issue in enclosed, mechanically stirred, and aerated composting facilities except at the receiving area where the incoming material may be stored prior to being fed into the machinery.

Composting operators have identified a large number of species of flies breeding in material handled at a composting operation. The common house fly, *Musca domestica*, is reported to predominate in many of the locations where identification occurred. Identifying the species of fly is relatively unimportant at an actual site because the control measures are essentially the same for all species. The life cycle of the house fly is typically between 7 and 14 days under favorable conditions. Control measures must interrupt this cycle before the emergence of the adult flies.

The initial handling of the feed material causes a large number of larvae to be killed. This is particularly important if grinding is involved to reduce the size of particles. After the composting material is placed in piles, windrows, or pits, the heating process kills the larvae in the center of the pile. During this heating process, however, many of the larvae move from the center to the outside of the pile where conditions are more favorable for survival. In some locations, larvae move from the pile to surrounding soil to escape the heat and later emerge as adults. This supports the design of manure composting facilities that provide an impervious surface of concrete or asphalt for the initial handling and active composting areas to facilitate fly control. During seasons when fly problems are particularly severe, more frequent turning may be necessary to prevent flies. Some operators have found daily turning to be necessary, particularly if the feed material arrives at the composting site with a high population of fly eggs and larvae.

Some operators have found that insulating the outer surface of the compost windrow with stabilized compost or other dry material allows the surface of the active compost to reach a sufficiently high temperature to control flies.

Whatever the material being composted and the location of the facility, fly control is an important consideration in both design and operation of a composting facility. The principles of good housekeeping are critical. Scattered compost and manure are particularly likely to attract flies because they are not producing heat. Frequent turning helps to control flies in the active piles or windrows. The finished compost being aged does not attract flies so it can be stored without creating problems. Judicious spraying of insecticides and the use of fly traps around a composting facility can supplement fly management, but the overall most effective program is one that stresses scrupulous housekeeping and astute compost management.

Quality of Compost

Decisions regarding when composting is complete and the quality of the compost require some experience. Most workers agree that active composting is complete when a pile with adequate moisture no longer heats upon turning. Composting also is considered complete when a pile cools from the exhaustion of available organic matter. At this stage, most operators move the compost to a storage area for maturing. If bags of compost are to be sold to retail consumers, it is most appropriate to grind the aged material and place it in bags immediately prior to sale to avoid bags splitting, rotting, or otherwise becoming unattractive in storage.

The nutrient composition of compost is highly variable because the final material is dependent upon the starting material. The initial carbon-to-nitrogen ratio is particularly important to the nitrogen content. If the ratio is too high, composting is slowed and the final product contains most of the initial nitrogen but in a mass that has been reduced by 30 to 50%. If the carbon-to-nitrogen ratio is less than 20:1, nitrogen loss due to ammonia volatilization can be expected. Phosphorus and potassium losses are negligible if rainfall does not leach them from the material. Table 9.1 provides information on the concentration range of various materials in finished compost.

Persons producing compost for sale on the consumer market have the opportunity to supplement the compost to better meet the needs of individual consumers. The relative concentrations of nitrogen, phosphorus, and potassium can easily be adjusted by adding appropriate amounts of chemical nutrients and still preserve the benefit of compost addition to enhance soil structure.

Table 9.1. Typical composition ranges of
finished compost from animal manures

Constituent	% by weight
Organic matter	40–60
Carbon	10–50
Nitrogen as N	0.4–3.5
Phosphorus as P_2O_5	0.3–3.5
Potassium as K_2O	0.5–1.8
Ash	20–50
Calcium as CaO	1.5–7

Manure Composting Facilities

Various physical arrangements have been used to facilitate the composting process. The process is essentially independent of the physical facility so the choice of facility design is largely an engineering issue based on
the size of the composting operation, land area available, local climate,
labor availability, and the preference of the operator. The open pile or
windrow arrangement is the most frequently used design because it requires a minimal investment and allows for convenient expansion or size
reduction in response to either market conditions or the availability of
feedstuffs. Figure 9.4 is a photograph of a manure composting system in
operation.

Open Pile or Windrow Composting

An open pile or windrow placed directly on the ground or on a paved
area is the most popular approach to composting on a worldwide basis.
Frequently, the windrow or pile is in a shallow pit for operator convenience. The exact design depends on local conditions such as temperature, rainfall, snow, and wind as well as on labor availability and cost. In
areas with extended rainy periods, paved areas certainly reduce operational difficulties. Drainage is important in the design of facilities with extended rainy seasons. Runoff from composting slabs should be collected
and managed much like the runoff from animal confinement areas.

Typically, the first step in composting is a sorting and size reduction
process. It is desirable to remove noncompostable debris that may be in
the waste supply. This step may be unnecessary if a clean source of animal or poultry waste is to be composted. Frequently, it is also necessary
to have a mixing operation if two or more different waste materials are
being combined as the feed material. Once prepared, the moisture content is adjusted if necessary and piles or windrows are prepared.

Figure 9.4.   Manure composting operation being used to produce a low odor-intensity soil amendment for use by home gardeners or around ornamental plants.

In forming compost piles or windrows, the material should be stacked loosely to allow as much air to be incorporated as possible. Windrows may be of any length that proves convenient and that is compatible with the site. Windrow or pile height is more important than length. If piles or windrows are too tall, the material at the bottom of the pile is compressed from its own weight. Pile and windrow depths of 4 to 5 feet are generally appropriate. If piles are too shallow, there is too much heat and moisture loss and the pile temperature is not sufficient to kill pathogens and weed seeds and to prevent fly propagation. It may be desirable to increase pile height during cold weather to conserve heat. Windrow height also may be restricted by the choice of equipment for stacking and mixing.

Pit Composting
Most pit composting of animal manure is done in rectangular cross-sectional pits whose width is selected based on the mechanical stirring equipment that rides along the top of the dividing walls. Typically, the mechanical stirring equipment is a horizontal-shafted rotating set of paddles that lift the composting material and place it farther down the pit. The paddles move relatively slowly through the compost and the device also moves slowly along the length of the pit. Alternatively, compost in

pits may be stirred by using a front-end loader to move the active compost from one pit to another. Except for the stirring mechanism, the considerations for pits are essentially the same as for windrow composting.

There are also pit or tank-type digesters with mechanical aeration and stirring equipment installed as integral parts of the composter package. These devices are generally sold as integrated packages and operated according to manufacturer's guidelines. They are most often selected where high rates of composting are required or where weather is too inclement for uncovered compost stacks.

Incorporating Compost Production in a Livestock Enterprise

Bedded Systems

There are several alternatives for incorporating compost production into livestock and poultry production schemes. Among the most obvious is the treatment of spent poultry litter from broiler production houses in which the birds are raised on a layer of litter. The litter is typically sawdust, wood shavings, straw, or other easily available material that is dry and has a sufficient water holding capacity to absorb droppings and maintain aerobic conditions beneath the birds. Litter may be replaced after each cycle of birds or it may be fluffed between flocks and used for up to a year. Upon removal from the building, the used litter may be immediately incorporated into a compost process. This creates a convenient batch process that is compatible with many operator's time schedules. After 30 to 45 days of active composting, the material is stockpiled for curing. After curing, it is typically processed through a grinder to achieve relatively uniform particle sizes and then bagged for marketing to the home gardener or for sale through a landscaping service.

Another obvious opportunity is the composting of bedded manure from horse stables or from calf-raising facilities. In both of these cases, the animals are typically maintained on bedding, straw, or wood shavings that are removed periodically. In many parts of the Unites States, this bedding material from horse stables is in considerable demand for compost production by mushroom producers who traditionally have used it as the starting material for growing edible mushrooms. Composted cow manure is of interest to organic garden enthusiasts seeking a natural source of plant nutrients and soil humus-building properties.

Scraped Manure Systems

There are a number of beef and dairy cattle confinement facilities in which manure is scraped from the housing area either with permanently installed scraper systems or with the use of tractor-mounted blades. Such manure contains too much water for composting as it is removed from the confinement building but can be mixed with a dry waste source to produce a material suitable for composting. Alternatively, in many areas, the manure can be allowed to air dry then be placed in windrows or piles for composting.

Water-carried Systems

Flushed manure transport systems are popular in many parts of the world for handling swine and dairy cattle manure. The flushed material typically has a total solids content between 1 and 2%. This material as removed is too dilute for incorporation into a compost operation. A common initial treatment step is to pass this flushed material over a solid–liquid separator to reduce the organic load on downstream biological treatment processes. The separated solids are suitable for composting. This material is used by ornamental nurseries as a mulch or in the formulating of their potting mixes to produce a highly friable medium for packing plant roots immediately prior to sale and for incorporation in the soil where these plants will be placed to achieve a soil into which the roots can more easily penetrate.

In-building Composting

There are a growing number of confinement systems in which the animals, most frequently pigs, are being raised on a surface covered with up to 3 feet of bedding material. This material provides a warm floor and the animal activity tends to incorporate the manure and urine into the bedding material. This then becomes a continuous composting operation in which the role of the manager is to monitor the moisture content of the bedding and to periodically stir the material to break up any surface compaction that may have occurred and to work the manure deeper into the composting mass. One requirement of a system of this type is that the watering system does not leak sufficiently to saturate the pit or bedding, causing it to become anaerobic. The bedding is periodically removed for use as a soil amendment and replaced with fresh bedding.

Dead Animal Composting

Dead animal disposal is one of the long-term challenges of confinement livestock and poultry producers. Dead animals that are not promptly removed from the site become a significant odor as well as aesthetic problem. The public is usually highly sensitive to an accumulation of dead animals. In addition, there is a significant public health concern involved in having dead animals improperly managed. They are subject to scavenging by wild animals and subsequently by neighbor's pets.

Prompt removal by a rendering service is an optimal solution to the dead animal problem, however, in many locations there is not a rendering plant within an economically accessible distance or the service may be unwilling to collect carcasses with a sufficient frequency to meet local health regulations. Local health regulations usually insist upon dead animal removal within 24 or 48 hours. Rendering operations also may be unwilling to accept dead poultry.

The traditional alternative has been to bury dead animals. Groundwater concerns have made this option less available in many areas; furthermore, soil conditions are not suitable for burial in many areas during a significant part of the year. Burial requires a high level of management if it is to be accomplished without nuisance conditions developing.

Incinerators are available for the disposal of dead animals but they are expensive to purchase, require fuel for operation, and represent a labor demand. Many incinerators require that the animal be macerated prior to incineration.

Procedures have evolved for the incorporation of dead animals, particularly poultry and more recently pigs, into an active compost bin. Generally, the compost bin is operated specifically for dead animal disposal. It is a carefully designed bin requiring a layer of sawdust or other dry absorbent material, some manure, appropriate moisture content, and then a layer of dead birds or a dead pig, followed by additional layers of manure and sawdust. The batch system composts for approximately 3 months without being disturbed. After 3 months, the compost should be moved to a second bin to affect mixing and reaeration. Reheating can be expected after this turning operation. After an additional 3 months, when this secondary bin is opened to remove the compost, all that remains of the dead animals is a collection of bone fragments. The compost may be moved to a curing pile or land-applied, or a portion may be incorporated into the next bin along with the dead animals. The elevated temperature of the composting material speeds decomposition of the incorporated animal and the moist, porous material traps the gaseous products of decomposition so that odors are held in check.

Depending upon the number of animals to be composted and local weather conditions, alternative composter designs have been used. The simplest involves the use of large hay bales to form three or more bins. One of the bins is used to store fresh sawdust or other carbonaceous material; the other two serve as primary and secondary composting bins. In areas where there is sufficient rainfall to cause the composting mass to become soggy, where a more controlled process is desired, or where

Figure 9.5.   Roofed composting facility designed to provide an alternative disposal method for dead animals that is particularly useful in areas where a commercial rendering service is not available.

there are more than 200 pounds of dead carcasses per week for composting, a roofed composter with concrete bin walls is more suitable and leads to a higher quality compost. The volume of composting bins required depends upon the number and size of carcasses to be composted. One sizing guideline is to provide 20 cubic feet of primary and secondary bin volume per pound of carcass composted daily. This would be equivalent to 1.5 cubic feet per pound of carcass generated per month; thus, if it is the local experience that 20 pounds of baby pig carcasses are generated weekly plus one 130-pound finishing pig each month, the total bin volume would be $1.5((20 \times 4) + 130)$ or 315 cubic feet. A roofed composter with concrete bin walls suggested by the University of Missouri Extension Service is similar to but smaller than the one shown in Figure 9.5.

Dead animal composting has proven to be an answer to the problem for many remotely located livestock and poultry operators. The process does require, however, that the bin be constructed according to detailed specifications to avoid flies and to protect the composting material from scavenging animals. The moisture content of the bin needs monitoring because either insufficient or excessive moisture leads to operational problems, odors, and flies, to which neighbors are likely to be sensitive. It is important that the operator of a dead animal composter understand how it is to function to avoid nuisance conditions.

10

Manure Application to Crop and Pastureland

Traditionally, manure from livestock and poultry has been viewed as a valuable source of plant nutrients needed by crop and pastureland to achieve greater productivity. Both the literature and experience support this view. Application of manure to crop and pastureland restores a portion of the nitrogen, phosphorus, potassium, and other trace nutrients removed from land when crops are harvested and taken off the land for use elsewhere. During the first half of the 20th century, the typical farm in the United States grew a variety of crops and fed a portion of these crops to livestock and poultry. Their manure was returned to the land. As long as animal manure was the only readily available source of replacement plant nutrients, it was regarded as valuable to the farmer and worthy of protection and conservation.

The arrival of the industrial age and the availability of chemical fertilizer, however, created an alternative source of nitrogen, phosphorus, and potassium. This chemical fertilizer was frequently more convenient to apply than manure and was of known nutritive value, hence more easily applied in proper amounts. Thus, animal and poultry manure became less valuable. The land application concept has remained, however, and even today the predominate mode of manure disposal in the United States and western Europe is application to crop and pastureland. The basic application logic has changed, however, from application to achieve economic benefit (improved crop yield) to one of safe and environmentally responsible disposal of the manure. This change does not mean that manure is no longer of value as a source of plant nutrients but that there are other sources of these nutrients that are frequently less expensive to use than manure when considering the cost of collection, storage, and application.

Manure disposal has become even more challenging in the past 20 years because livestock and poultry production has become more concentrated in larger enterprises and crop production and animal feeding have become less directly connected. Because grain producers have become more remotely located from the livestock producers, the ability to return manure to the land from which the grain was produced has become impossible. Thus, the U.S. livestock producer faces the dilemma of frequently having more manure available than the crops grown on his or her farm can use, and in many locations, more manure than all the crop producers in the area can use. Thus, the concentration of manure in a specific area is frequently in excess of that which can be used within the economic transport distance. Manure management becomes more critical under these conditions because the manure has negative value. Suddenly, the designer finds his or her role dominated by the goals of environmental protection rather than the goal of resource management.

Manure application to crop and pastureland is addressed in this chapter as a means of improving crop production and as a means of preserving environmental quality. The location, size, and land holdings of the enterprise determine which goal predominates in a particular location.

Alternate Manure Application Strategies

There are two possible questions that can be raised concerning the application of manure to crop and pastureland. One is, How much manure must be added to achieve the potential crop yield of a particular field under the proposed management plan? Asking this question implies that the manure has value and if not applied to this particular field, would be applied to another where it also would have value. Under this strategy, application rates would logically be addressed by an agronomist based on appropriate soil sampling, knowledge of the soil, information on the anticipated yield of the crop in the area under consideration, and an analysis of the manure to be applied. This approach represents a high level of nutrient management toward the goal of "prescription farming" in which manure is considered as an input to be minimized for greatest profit. Nitrogen, phosphorus, and potassium in particular would be added to replace that used by the crop. Other inputs such as irrigation water, pesticides, and cultural practices would be similarly balanced to provide the greatest long-term return to the landowner for his or her investment and labor.

The second question that can be used to determine manure application rates is, How much manure can be applied without causing measurable damage to the environmental quality of the area? This question acknowledges that if excessive amounts of manure are applied to crop or pastureland, the soil resource is damaged and yields are lowered, odors are created that are carried to neighboring residents, nitrogen not used by the crop is transported to the groundwater, and runoff from the area

Table 10.1.   Nutrient removal capabilities of various crops as published by the University of Missouri Extension Service and the Missouri Department of Natural Resources

Crop	Yield/acre	Part harvested	N, lb	P, lb	K, lb
Reed canary grass	6.1 ton	All	359	35	299
Fescue, grass for hay	5.0 ton	All	275	26	269
Alfalfa for hay	4.0 ton	All	196	19	156
Orchard grass for hay	2.5 ton	All	65	9	95
Timothy for hay	2.5 ton	All	53	7	79
Bluegrass for hay	2.0 ton	All	52	11	13
Corn	80 bu	Grain only	64	13	13
Corn	80 bu	All	115	18	96
Corn for silage	150 bu.	All	185	35	178
Milo	70 bu	Grain only	61	10	12
Milo	70 bu	All	81	14	63

Notes: 1) Reed canary grass, although it has the potential to remove large amounts of N, is also among the most sensitive grass to salt concentrations. Application of liquid manure with a high electrical conductivity may reduce yield. 2) Alfalfa will normally fix most of the required N from the air. Because N is applied as manure, more of the applied N is used and less is biologically fixed. 3) Corn grown for silage under optimal agronomic conditions can use the amounts of N, P, and K noted. If the crop is planted late, receives inadequate water, or suffers other yield reductions, the nutrient use rates shown are not achieved.

carries more phosphorus and potentially disease-causing microorganisms to the receiving stream. Examples exist in which each of these damages has occurred. This question assumes that manure is a waste material to be applied to land at the maximum acceptable rate based upon ability of regulators or off-site neighbors to detect damage to them or their property. One of the challenging aspects of this approach is that several of the potential damages from excessive manure application rates are not easily detected immediately after application. The time required for excessive nitrogen applications to reach the groundwater aquifer and be transported to a neighboring well may be years. Thus, if we wait for such evidence to be gathered, vast areas may be severely damaged in terms of future productivity and livability.

Because neither of the above-mentioned questions is easily answered, manure application rates are most commonly based on a consideration of the amount of nitrogen the crop can be expected to use. Nitrogen use rates are widely available and have been adjusted for the climatic conditions that exist in various locations. Thus, a landowner can determine a recommended nitrogen application rate for each crop. This rate is frequently used as the starting point for calculating manure application rates. Table 10.1 shows an example of nutrient use rate information that can be used to calculate manure application rates. In some regions of the United States where surface water pollution of lakes and natural wetlands has become a concern, phosphorus application rates are limited. In these areas, manure application rates are based on the allowable amount of phosphorus that can be applied to an area.

As Table 10.1 shows, nutrient uptake rates of crops vary widely. Uptake rates not only vary among crops but also are dependent upon the weather during a particular growing season and the extent to which the farmer is able to manage tillage, water supply, and the weather to achieve the full genetic potential of the crop. Compounding the uncertainties of nutrient use is the timing of nutrient uptake. Crops do not grow uniformly throughout the year and therefore do not use the nutrients at a uniform rate. It is appropriate to consider timing in planning a manure application schedule. Applying a full year's nitrogen application in the fall after the crop has been harvested would not be a good idea in those parts of the country with winter rainfall and highly permeable soils because much of the nitrogen would be transported through the soil profile beyond the root zone before it could be used. Similarly, the application of a full season's supply of nutrients to a highly permeable soil as a single application would promote loss of nitrogen to the groundwater.

Hazards of Excess Manure Application

Manure, when applied to crop and pastureland in appropriate amounts with appropriate techniques, serves as a source of plant nutrients and humus. The plant nutrients replace nitrogen, phosphorus, potassium, and trace nutrients that are necessary for crop growth. When one or more of these nutrients become limiting, crop yields are reduced and the farming operation is less productive. The organic matter of manure or compost is beneficial in maintaining soil structure. Soil containing appropriate levels of organic matter has a higher infiltration rate, conducts water to the plant roots more effectively, and displays an enhanced quality known as tilth. Tilth is that general quality of a soil reflecting its suitability for cultivation and seedbed formation. Soils that lack adequate tilth are generally less productive, have a lower rate of water movement, and are more subject to erosion damage. Unfortunately, there is no widely accepted quantitative measure of soil tilth so this aspect of organic fertilization can only be discussed in qualitative terms.

Excess Nitrogen Application

When fresh or stored manure is applied to land, most of the nitrogen is present as ammonia in solution or as organic nitrogen. The ammonia fraction tends to be absorbed on soil particles because of the net positive charge of ammonium. A portion of the ammonia is subject to volatilization, however, so manure should be incorporated into the soil as quickly as possible to avoid this loss. Prompt incorporation also reduces odor complaints. The organic material containing nitrogen is relatively stable and releases nitrogen over a period of up to 2 years, acting as a slow-release fertilizer. This slow-release characteristic is the basis for a common practice of allowing a slight excess of nitrogen to be applied to a field not previously fertilized with manure. A portion of the organic nitrogen applied is not available for crop use during the year it is applied.

The ammonia that is applied may be 50 to 90% of the total nitrogen, depending upon the way the manure has been stored or treated. Fresh manure has less of the nitrogen in the ammonia form than an anaerobic lagoon. The applied ammonia oxidizes to nitrate relatively quickly after application to land. Aerobic bacteria capable of this energy-releasing transformation are widely available in the soil. Once converted to nitrate, the nitrogen is totally soluble in water and no longer attracted to soil particles, so it is free to move downward with percolating water. This water as it carries nitrate is the source of nutrients to the plant roots. If excess nitrate is present, beyond that which can be used by the crop, it tends to be carried beyond the root zone, hence becoming unavailable to the crop. Nitrogen that moves beyond the root zone is then en route to the groundwater aquifer. Excess water application whether due to rainfall or excess irrigation has the same effect, namely, moving the nitrate through the soil profile before the plant roots have an opportunity to use it.

A small nitrogen loss to the groundwater is part of the natural process. Naturally occurring soil organic material decomposes and that which is not incorporated into plant tissue is available for leaching. When excess manure or commercial fertilizer is added to land, the amount of nitrogen moving to the groundwater increases dramatically. Under natural conditions, nitrate concentrations in groundwater are generally less than 5 milligrams per liter expressed as nitrogen, often less than 1 milligram per liter. Groundwater with more than 5 milligrams nitrate per liter is generally indicative of previous high nitrogen applications either due to overfertilization or to improper manure handling techniques. There are also areas of naturally high nitrate concentrations in groundwater but these areas are unusual. Nitrate concentrations of more than 200 milligrams per liter have been measured in shallow wells located adjacent to heavily fertilized fields, to fields to which manure has been applied in excess of plant uptake, and to previous manure disposal areas.

The nitrate problem due to excess manure or fertilizer application is made more serious because of the slow rate of groundwater movement and the difficulties inherent in groundwater monitoring. Thus, it may be several years or even decades of excess application before the difficulty is discovered.

Nitrate is of concern in drinking water because of the danger it poses to infants and pregnant women. The digestive systems of infants less than 6 months of age, and unborn fetuses can reduce nitrate consumed or passed from the mother into nitrite. This nitrite then enters the bloodstream and reacts with the hemoglobin to replace the oxygen that is normally transported throughout the body. The infant then suffers oxygen deprivation, hence the common name, blue baby syndrome. Officially called methemoglobinemia, this condition can be fatal if not detected in time. In response to this possibility, the drinking water standards in the United States specify the nitrate concentration should be less than 10 milligrams per liter expressed as nitrogen. Higher concentrations are generally not hazardous to older children or adults. There is evidence that high nitrate concentrations are harmful to livestock. Symptoms generally are poor efficiency and increased rates of abortions and stillbirths.

Nitrate is among the most difficult ion to remove from water because of its solubility. Boiling is typically suggested to overcome microbiological hazards, but it only worsens the nitrate problem by further concentrating the ion. Other disinfection techniques are equally ineffective. Distillation and some of the synthetic ion exchange materials are effective.

Excess Phosphorus Application

In contrast to nitrogen, which is primarily a groundwater pollutant, phosphorus is most frequently of concern in the enrichment of surface waters. Algae and higher aquatic plants tend to grow in lakes and reservoirs in response to sunlight, water temperature, and nutrients. In shallow waters, particularly those in warmer climates, the abundance of algae is controlled because one of the essential nutrients becomes limiting. Phosphorus is frequently the nutrient of choice to be used. To effectively limit algal production, the concentration of phosphorus needs to be 0.05 milligrams per liter or less.

Manure application rates are generally calculated to achieve a nitrogen application rate that balances the uptake of the crop. Because of the nitrogen losses during storage and treatment, this strategy frequently results in the application of phosphorus at rates much higher than required by the crops or that can be removed by the crops. Traditional wisdom has held that phosphorus remains bound to the soil particles and with good erosion control practices, does not run off into surface waters or percolate into groundwater. Recent studies, however, have shown that phosphorus is more mobile than previously thought. This suggests that the practice of overapplication of phosphorus on a long-term basis contributes more phosphorus to surface waters than previously thought. Groundwater entering surface streams in agricultural areas has been found to contain approximately 0.20 milligrams phosphorus per liter, which is well above the suggested concentration of 0.05 milligrams per liter. Water and soil quality concerns have led to the proposal of national standards and adoption of regional criteria that manure applications must not exceed agronomic phosphorus requirements.

Should phosphorus application rates be mandated to match crop use rates, it would have a major impact on livestock and poultry producers. It would increase the land area required for manure disposal by up to four times for many commercial livestock producers and require other farmers to supplement their manure application with chemical sources of nitrogen. In fact, advanced nutritional strategies show great potential for reducing the nitrogen and phosphorus content, amount, and odor of manure in the future.

There are other possibilities for reducing the application of phosphorus from manure. One is by adjusting the feed ration to effectively reduce phosphorus intake, therefore the amount of phosphorus in manure. The addition of phytase enzyme to animal feed allows the phosphorus concentration to be reduced by increasing the animal use of phosphorus in the ration.

Determining How Much Manure to Apply

Manure whether solid or liquid, fresh or treated, or concentrated or dilute is a potential water pollutant and therefore deserves to be handled with care. In addition, it has value as a fertilizer and can therefore be of value to the crop or pastureland to which it is applied. There are several alternatives to be considered in deciding the amount of manure to apply at a particular time to a particular field.

Basis of Application

Nitrogen is the nutrient of greatest concern to groundwater pollution. If excess nitrogen is applied, beyond the amount that can be used by the crop, the excess is converted to mobile nitrate that is transported beyond the root zone by water moving from the surface toward the water table.

Nitrogen as a basis. Anticipated nitrogen use is a frequent basis for calculating the amount to be applied. Estimating the amount of nitrogen needed is most accurate when based on a soil analysis that determines the amount of residual nitrogen available in the soil profile. This information allows a determination of how much additional nitrogen to apply. Whether it is appropriate to apply a full season's nitrogen fertilization in a single application is best determined on a local basis. If the area under consideration receives sufficient rainfall during the growing season to move nitrogen beyond the root zone, the application should be split so the nitrogen needed later in the growing season can be applied when needed. The amount of nitrogen that can be used also depends upon the yield that can reasonably be anticipated. A competent agronomist familiar with local crop yields can, based on the soil analysis, estimate the amount of nitrogen that should be applied.

Phosphorus as a basis. Nitrogen alone is not always the appropriate basis for manure application. There is a growing realization that if manure is applied according to the nitrogen needs of the crop, application of most types of manure results in an excess of phosphorus being applied. Considerable quantities of phosphorus are bound with the soil so the total concentration is in excess of that measured by standard analyses to determine the amount of phosphorus available to establish seedlings. Thus, many have argued that excess phosphorus is not a problem in agricultural soils because it is quickly bound to soil in the profile and is not mobile. The alternate logic is that phosphorus is bound to the surface soil and because an excess is applied, the excess contributes to phosphorus enrichment of surface waters in the area that receives runoff from fields to which excess phosphorus has been added. Based on this concern, areas of the country where eutrophication of surface waters is contributing to undesirable ecological changes have begun to place limits on how much phosphorus can be applied per acre. Economics would dictate that if there is more land to be fertilized than there is manure available, basing application rates on phosphorus rather than nitrogen is the more intelligent strategy. In these cases and where there is a concern over excess phosphorus in the soil, it is appropriate to consider basing the manure application rate on the agronomic requirement for phosphorus. If phosphorus is used as the basis, it is important to calculate the resulting nitrogen application rate and where necessary, supplement the manure with the application of an alternate nitrogen-containing material.

Whether the nutrient application rate is based on nitrogen or phosphorus, it is important to know the concentration of these nutrients in the material to be applied. These concentrations are best determined by laboratory analyses. The alternative is to estimate the nutrient concentrations based upon the dilution, storage, and treatment processes used. The estimating process is subject to serious errors and is not an appropriate technique for precision application.

Nitrogen loss during application. In planning for the application of livestock and poultry manure to meet the nutrient needs of a crop or pasture, it is important to include an allowance for the amount of nitrogen that is lost between the storage or treatment site and the soil profile. Nitrogen may be lost due to volatilization as ammonia because lagoon or other liquid effluents are sprayed through a nozzle and travel through the air, or as solid manure lies on the ground after spreading and before

Table 10.2. Estimated nitrogen losses during the application of manure to crop or pastureland by using various application techniques

Application technique	Estimated % nitrogen lost during application
Hauling solid manure and bedding followed by spreading on bare cropland with a manure spreader, incorporated within 24 hours	5
Hauling solid manure and bedding followed by spreading on cropland or vegetation-covered pasture with a manure spreader, not incorporated	25
Hauling liquid manure slurry in an enclosed tank truck or wagon from an anaerobic storage tank followed by injection directly into the soil	2
Pumping liquid or slurry manure in a pressure pipeline from an anaerobic storage tank or lagoon followed by pressurized injection directly into the soil	3
Hauling liquid manure slurry in an enclosed tank truck or wagon from an anaerobic storage tank followed by spray distribution on the soil or vegetation-covered surface, no incorporation	20
Hauling liquid manure slurry in an enclosed tank truck or wagon from an anaerobic storage tank followed by spray distribution on the soil or vegetation-covered surface, incorporation within 24 hours	5
Pumping liquid manure from an anaerobic storage or a stored slurry by using a pump, pipeline, and big gun nozzle under high pressure to get maximum distribution and uniformity	30
Pumping liquid manure from an anaerobic lagoon, storage basin, or a stored slurry by using a pump, pipeline, and a series of small-diameter nozzles under high pressure to get maximum distribution and uniformity	30
Pumping liquid manure from an anaerobic lagoon, storage basin, or a stored slurry by using a pump, pipeline, and a series of low-pressure nozzles designed to create larger droplets	20

incorporation. During high winds, significant amounts of nitrogen are transported off-site as wind-carried aerosols. Large amounts of ammonia also tend to be volatilized if the wastewater has a pH above 8.0 and the weather is warm. Table 10.2 provides estimates of the percentage of nitrogen loss under various systems of manure application. The values presented represent best estimates of what can be anticipated but are wisely adjusted based on local experience, sampling, or knowledge of conditions that would cause either more or less nitrogen to be lost.

It is frequently desirable to determine how much nitrogen is being applied rather than to depend upon estimated losses as shown in

Table 10.2. If the wastewater is being applied with irrigation equipment, the usual way is to place a series of collection cans at locations throughout the spray field and catch samples of water similar to that entering the soil profile. The nitrogen concentration in the composite of the samples can be compared with the material being pumped and an on-site nitrogen loss calculated. The value measured in this way can be adjusted for weather and waste quality changes on an as-needed basis.

Calculating manure application volume. The amount of manure to be applied at a particular time, whether solid manure and bedding directly from the confinement area or liquid effluent from a storage or treatment device, is best based on knowing the nitrogen and phosphorus concentrations of the material available and the quantity of each of these nutrients that is appropriately applied at the time. For example, if the annual application rate to produce a corn crop in a particular area is 180 pounds of nitrogen and 60 pounds of phosphorus per acre, and these nutrients are best used by applying 50% in April before planting and 25% again in June and late July, the application volume for April would be calculated as follows:

Assuming Lagoon Effluent Applied with An Irrigation System

Based upon nitrogen application:

Amount of nitrogen desired in soil: 90 pounds per acre
Estimated nitrogen loss during application from Table 10.1: N_{loss}%
Amount of nitrogen to be pumped: $90/(100 - N_{loss})(0.01)$
Concentration of nitrogen in the lagoon effluent: C_N, milligrams per liter
Amount of nitrogen per acre-inch of lagoon effluent: pounds per acre-inch
   43,560 square feet per acre $(62.4/12)(C_N)(10^{-6})$
Depth of effluent to be applied: inches
   Amount of nitrogen to be pumped/nitrogen per inch of lagoon effluent

Using these values,
Amount of nitrogen desired in soil: 90 pounds per acre
Estimated nitrogen loss during application from Table 10.1: N_{loss} 25%

(Continued)

> Amount of nitrogen to be pumped: $90/(100 - 25)(0.01)$ = 120 pounds per acre
> Concentration of nitrogen in the lagoon effluent: 625 milligrams per liter
> Amount of nitrogen per acre-inch of lagoon effluent: pounds per acre-inch
>> 43,560 square feet per acre $(62.4/12)(625)(10^{-6})$ = 141.6 pounds per acre-inch
> Depth of effluent to be applied: inches
>> Amount of nitrogen to be pumped/nitrogen per inch of lagoon effluent
>>> $120/141.6 = 0.85$ inches

Based on Phosphorus Application

Amount of phosphorus to be applied: 30 pounds per acre
No phosphorus loss during application
Amount of phosphorus to be pumped, 30 pounds
Concentration of phosphorus in lagoon effluent, C_P, milligrams per liter
Amount of phosphorus per acre-inch of lagoon effluent: pounds per inch
> 43,560 square feet per acre $(62.4/12)(C_P)(10^{-6})$
Depth of effluent to be applied: inches
> Amount of phosphorus to be pumped/phosphorus per inch of lagoon effluent

Using these values,
Amount of phosphorus desired in soil: 30 pounds per acre
Concentration of phosphorus in the lagoon effluent: 175 milligrams per liter
Amount of phosphorus per acre-inch of lagoon effluent: pounds per acre-inch
> 43,560 square feet per acre $(62.4/12)(175)(10^{-6})$ = 39.6 pounds per acre-inch
Depth of effluent to be applied: inches
> Amount of phosphorus to be pumped/phosphorus per inch of lagoon effluent
>> $30/39.6 = 0.76$ inches

Assuming Liquid Slurry Being Soil Injected

Amount of nitrogen desired in soil: 90 pounds per acre
Estimated nitrogen loss during application from Table 10.1:
N_{loss}%
Amount of nitrogen to be hauled: $90/(100 - N_{loss})(0.01)$
Concentration of nitrogen in the slurry: C_N, pounds per 1,000
gallons
Amount of nitrogen per 1,500-gallon load: pounds per load
1,500 gallons $(1/1,000)(C_N)$
Loads of manure to apply per acre
Amount of nitrogen to be hauled/nitrogen per load of
manure

Using these values,
Amount of nitrogen desired in soil: 90 pounds per acre
Estimated nitrogen loss during application from Table 10.1:
N_{loss} 2%
Amount of nitrogen to be pumped: $90/(100 - 2)(0.01) = 92$
pounds per acre
Concentration of nitrogen in the slurry: 14 pounds per 1,000
gallons
Amount of nitrogen per 1,500-gallon load:
1,500 gallons $(1/1,000)$ (14 pounds/1,000 gallons) =
21 pounds
Loads of manure to apply per acre
92 pounds per acre/21 pounds per load = 4.4 loads per acre

Determining When and Where to Apply Manure

Several factors deserve consideration when establishing a schedule for manure application to crop or pastureland. Each field or crop may deserve individual consideration to provide maximum protection to the surrounding environment and to reap the maximum benefit from the manure. The goal is to apply manure so that the nutrients are used by the crop and so that both surface runoff and groundwater infiltration are minimized.

Where to Apply Manure

Manure is appropriate for application to lands upon which a crop is being grown that can use the nutrients and from which both surface runoff and groundwater infiltration can be controlled. This means the soil should be of sufficient quality to maintain a crop. The permeability of the soil should be sufficient to allow water to enter in adequate quantity to meet the water needs of the crop for which manure is applied. The permeability should not be so great, however, to preclude adequate water and nutrient retention to support the crop. Soils of high clay content may have such low infiltration rates that manure application is risky; similarly soils that are high in sand and gravel may not store water and water-soluble nutrients in the profile long enough to allow either to be used.

Slope is an important variable in selecting manure application areas. As the slope of a field increases, the application rate must be reduced to prevent overland flow of the manure. Surface irrigation techniques such as furrow irrigation with gated pipes or border irrigation can be used if the slope is less than 1%. Sprinkler systems can be used on land with slopes up to 5%. If the slope is greater than 5%, the application rate needs to be reduced from that indicated by the infiltration rate to prevent overland flow. Fields with extensive areas having a slope in excess of 10% are probably unacceptable for manure disposal other than by light application in support of pasture use.

Manure application sites also must be selected to avoid accidental problems that could produce long-term damage to the enterprise. For example, manure should not be applied to land within 100 feet of a stream. Greater distances are required from recreational water bodies or from public water supplies. Table 10.3 provides a list of considerations related to manure application areas that are provided to help avoid both environmental pollution and conflicts with alternate land uses.

When to Apply Manure

Each manure management system is designed to suggest a schedule for application to crop or pastureland. For example, if manure slurry is scraped to a storage tank, the size of the tank dictates the frequency of application. The size of the storage tank also determines the maximum time between applications. In areas of severe winter weather, it is typical to design storages to allow up to 6 months between the last application in the fall and the first in the spring. An experienced land manager can identify the earliest date at which manure can be beneficially applied. In addition, he or she can identify an autumn date beyond which it would be inappropriate to spread manure based on cold or wet weather. Thus, two application dates may be identified. Manure application dates during the growing season are ideally based on the crop. If hay is being grown with multiple cuttings, it is beneficial to schedule a light application after each cutting to stimulate regrowth and to capture the nutrients. Under such a condition, the full storage capacity of the tank would be used for the winter storage period but only a portion would be used during the summer.

Table 10.3. Factors involved in selecting a manure application site

Factor	Desirable qualities	Reasons for concern or rejection
Suitability of the site for crop production	Past record of high productivity, easily farmed	Poor productivity, difficult to manage
Soil type	Loam or sandy loam with moderate infiltration rate and good water holding capacity	Gravel, sand, fractured limestone or high clay content; infiltration rate either excessively high or low
Soil depth	Greater than 36 in.	Less than 18 in. over impermeable layer or fractured bedrock
Field slope	Less than 2%	Greater than 10%
Erodability	Low potential to erode; similar fields do not need terraces	High potential to erode; requires terracing or too steep to terrace
Relationship to watershed	Gently slopes into pasture or other permanently vegetated area	Slopes to stream, fishing or recreational pond, or well for public or household water supply
Proximity to public areas or neighboring residences	More than a mile from neighbors, schools, churches, commercial enterprises, or other sensitive land uses	Within 0.25 mile of neighbors or other sensitive land uses
Seasonality	Crop or grass cover at an actively growing stage in which nutrients can be used	Soil is frozen, snow-covered, water-saturated, or otherwise unable to accept additional water or when the soil surface is devoid of vegetation
Proximity to source of manure	Within 0.25 mile of manure-generation site or storage area	More than 0.5 mile away

It would be risky to not apply manure during the 6 months of the summer growing season. A heavy manure application late in the fall is not going to be beneficial to that season's crop and is subject to being leached away before the crops of the next season are established with a sufficient root system to reach the nitrogen. For example, a waste application schedule for an established grass field from which hay is harvested during the summer might be similar to that outlined in Table 10.4.

A runoff retention basin servicing an unroofed confinement facility in an arid region would have a very different wastewater application schedule. Under this condition, wastewater accumulates in the retention basin only as a result of runoff from the confinement area during and following rainfall. If the runoff retention basin is designed to have the minimum volume to capture runoff from the design storm, 25-year, 24-hour

Table 10.4. Example of a wastewater application and harvesting schedule for a combination hay field–pasture that is established

Time period	Activity
January 1–March 31	Store waste material because the crop is not growing and there is no nutrient use; there is likely a high risk of runoff due to frozen or snow-covered land
April 1–April 30	Apply wastewater to stimulate early season growth; exact date of application based on soil and weather conditions
May 1–June 10	Store wastewater while the crop is growing
June 10–July 1	Apply wastewater following harvest of the first hay crop
July 1–August 20	Store wastewater while the crop is growing.
August 20–September 1	Apply wastewater following harvest of the second hay crop
September 1–October 15	Store wastewater while the crop is growing
October 15–November 10	Apply wastewater following harvest of the third cutting of the hay crop; this application will be weather-dependent but the goals are to promote late-season growth for potential winter grazing and to use the remaining manure
November 19–December 31	Store wastewater

Note: This table represents a possible manure application schedule for a combination hay field–winter pasture in moderate climate. It needs to be adjusted to meet local conditions.

storm, land application is necessary as soon as the disposal site has dried sufficiently to accept additional liquid. Thus, with this design, the manager has very little discretion as to when land application is to be done. A more flexible design is to have a larger runoff retention basin so that some space can be allocated to storage and accumulation of the runoff from small storms. The manager then has the option of applying runoff from the storage volume in response to the need for the nutrients on the disposal area. The amount of additional volume is typically selected to hold the amount of runoff that can be anticipated during a typical winter. Under this condition, irrigation can be scheduled to make best use of the nutrients and water. In addition, it avoids the necessity of operating the irrigation system for less than a full pumping day.

Transport and Distribution of Manure

There are several alternatives available for transporting manure from the collection or storage site to the fields where it is to be applied. Of paramount importance in selecting a transport system is the form of the manure, i.e., solid, slurry, or liquid. Solid manure is that which is scraped from the surface of an outdoor feeding area during dry weather or from a bedded confinement facility. This manure can be stacked and because it does not have a liquid fraction, it drains from a pile. Solid manure typically has a moisture content between 25 and 75%. Solid manure may need to be transported from the point of production to a composting or alternate treatment site in addition to crop or pastureland.

Manure slurries contain too much water to be handled with solids handling equipment and too little water to be conveniently handled as liquids. Manure scraped from beneath slotted floors contains the feces, urine, and spilled water of the animals. Alleyways in confinement dairy barns are frequently designed to be scraped either with permanently installed scrapers or with the use of tractors fitted with a blade. This material generally does not flow on its own but must be pushed. Solids content is typically between 10 and 25%. The bottom fraction from a settling tank generally behaves as a slurry. Cattle feedlots occasionally yield a manure slurry when snow that has become incorporated in the manure melts sufficiently to flow.

Liquid manure, sometimes called wastewater, may arise as the runoff from rainfall landing on a manure-covered surface, or from the flushing of manure from a building with fresh or recycled water. Depending upon the source of the liquid manure, it may contain solids that are sufficiently large to interfere with conventional water handling equipment or it may be essentially free of large solids such as the effluent from a solid–liquid separating device or anaerobic lagoon liquid. Liquid manure contains less than 10% solids and frequently, if gutters are flushed, may contain from 0.5 to 1%. Table 10.5 provides an overview of the advantages and alternatives associated with the handling of the three alternate physical forms of manure. It is important to note that successful systems exist in which each of the forms is being managed.

Solid Manure Handling and Application

It is frequently possible to use direct hauling in the application of solid manure to cropland. For example, in the cleaning of beef cattle confinement pens, the manure on the feedlot surface is frequently scraped from the pen surface and loaded directly into a truck or wagon for hauling to the field. The high-use areas such as the apron along the feedbunk and the area around the watering device may require more frequent cleaning. Another location requiring more frequent cleaning may be along fence lines, particularly those at the lowest elevation where manure accumulates due to animal traffic. Cattle feedlot cleaning requires special care to

Table 10.5.   Alternate physical forms of manure and the associated handling possibilities, limitations, and frequently used applications

Physical form	Application	Advantages	Limitations
Solid manure	Smaller dairy or horse operations in which there is labor available to place and remove bedding Outdoor beef or dairy cattle confinement areas Poultry raising schemes, especially broiler production in which the birds are reared on a layer of dry mulch	Can result in a high degree of animal comfort particularly during cold weather Produces a material that has relatively less odor than either liquid or slurry manure Can be stored in a pile, with or without covering, depending upon local conditions and intended use Is suitable for composting with or without the addition of an additional carbon source	Generally results in more handling labor than the other alternatives Added bedding materials may be of limited availability or expensive to obtain, store, and distribute May produce a larger volume of manure than if a slurry system were used Nitrogen losses may be significant if the material is exposed to the weather for an extended period such as would be the case with feedlot solids
Slurry manure	Dairy operations on a solid floor in which manure is scraped one to three times daily Swine or beef cattle confinement facilities in which the animals are on slotted floors and manure either scraped daily or stored in pits beneath the floor	Produces the smallest volume of manure of the three alternatives Is compatible with a frequent application to cropland scheme to minimize nitrogen losses during treatment and storage Can be incorporated into a variety of solid-based nutrient recovery schemes such as composting, biogas production, or refeeding	Scraping systems, whether automatic or manual, may leave sufficient manure on the surface to contribute to an ongoing odor problem Handling equipment, especially pumping and scraping equipment, is more expensive and prone to mechanical breakdown than the liquid handling alternatives

(continued)

Table 10.5.   *(continued)*

Physical form	Application	Advantages	Limitations
Slurry manure (continued)	An operation proposing to treat the manure with anaerobic digestion and biogas recovery for energy production		
Liquid manure	Flushed dairy, swine, or beef confinement buildings may be flushed one to three times daily Flushed manure can be passed over or through a solid–liquid separator yielding two product streams Anaerobic lagoon liquid or runoff retention basin contents can be applied to cropland as a liquid	Liquid manure handling can be automated using equipment available for standard irrigation Liquid manure can be mixed with other irrigation water and applied to crop or pastureland with minimal additional equipment Involves the greatest ammonia release of the three alternatives	Requires the availability of a large water supply or alternatively a sufficient treatment system to allow recycling Initial equipment investment is higher than the other two alternatives Where spraying is the method of land application, ammonia and odor release may become a significant concern

avoid removal of the compacted soil layer that develops due to animal hoof action. This layer is important because its impermeable nature reduces or eliminates downward percolation of contaminants.

If solid manure is to be loaded frequently from a surfaced area, a reinforced concrete wall against which to push the manure is helpful. Another alternative is to have a ramp from which manure can be pushed so it falls directly into the spreader or truck.

Storage of solid manure. There are several benefits of storing solid manure if it is to be accumulated on a continuous basis or if manure-covered surfaces are to be cleaned at times in which it is either inconvenient or inappropriate to haul manure. The most common reason to store solid manure is to avoid application on muddy, frozen, or snow-covered fields where the field might be damaged by heavy vehicle traffic or there is a high potential for runoff. Storage also provides greater flexibility to the operator, allowing manure to be spread when it has greatest value for crop production or when it is more convenient. There is also an improved efficiency involved in hauling manure at a few selected times rather than continuously.

The solid fraction from a solid–liquid separator is frequently stored as a pile near the separator and moved with a front-end loader. The pad on which the separated solids are stored should be surfaced with an impervious material. Concrete is the most common choice. In the design of a storage pad, provision must be included for the handling of any drainage that flows from the pile and any rainfall. Runoff from the solids storage pad may be diverted to the lagoon or liquid storage tank if either of these are part of the system. If there is no liquid collection system, the material may be collected in a separate tank or the facility roofed to minimize the quantity involved. Roofed solids storage areas are common in high-rainfall areas.

In the design of solid manure storages, it is important to consider the presence of any high water table problems that might be possible in the area. It is suggested that the bottom of a manure storage be at least 3 feet above the maximum water table elevation. If a below-grade floor is being considered for solid manure storage, access should be considered. Steep ramps are a major source of frustration during wet or cold weather.

The manure storage facility should be located with paramount consideration of the convenience of removing the manure from the confinement area. Convenient access for the equipment to haul the manure out is also important. Because the manure storage unit can be a source of odors, particularly during unloading times, it is appropriate that some thought be given to its location relative to neighbors or to other odor receptors of concern.

Although most solid manure is moved by scrapers or front-end loaders, there are also mechanical conveyors and piston-type pumps that can be used to move solid manure with or without bedding. In piston pump installations, manure is scraped into a below-floor hopper and fed into an 8- to 10-inch reciprocating plunger that forces manure through a pipe to a remote storage location.

Transporting solid manure. Solid manure is typically transported in a manure spreader or truck fitted with an unloading and spreading device similar to that shown in Figure 10.1, which allows the manure to be distributed as the vehicle is driven or pulled through the field. Spreading widths of 5 to 8 feet are typical. Loads vary between 2 and 5 tons. Table

Figure 10.1.   Typical manure spreader used for the transport and distribution of solid manure on crop or pastureland

10.6 provides the manure application rate in tons per acre for various sizes of loads spread over distances from 400 to 2,000 feet. If the distance between passes is different from 6 feet, the application rate can be determined by multiplying the application rates from the table by a factor of 6/actual distance between successive passes.

Slurry Manure Handling and Transport

Manure slurries typically contain from 5 to 15% solids and may be fresh or may be stored prior to application to cropland. One method of handling manure slurries is the use of a ramp to load it directly into a truck or tank wagon as shown in Figure 10.2. Among the advantages of slurries is that the volume of material to be transported is less than if the manure were diluted to a more liquid form. Equipment is available for the storage, agitation, pumping, transport, spreading, and injection of manure slurries. Equipment for the handling of manure slurries is typically more expensive than a similar device for liquid manure because a more rugged unit is required to accept the higher pressures and abrasive slurry.

Storage of manure slurries. Manure slurries arise in slotted-floor buildings and under these conditions, it is typical to store the manure as a slurry in the pit directly beneath the slotted-floor area (see Chapter 5). The

Table 10.6.   Application rates of solid manure, tons per acre for various sizes of loads spread over travel distances from 400 to 2,000 feet. Based on a 6-foot distance between passes and a spreader traveling speed of 3 miles per hour

Unloading time in field, min.	Distance of travel, ft	Load size		
		3 tons	4 tons	5 tons
1.5	400	54	73	91
2.5	600	36	48	61
3	800	27	36	45
3.5	1,000	22	29	36
5.0	1,300	17	22	28
6.0	1,600	14	18	23
7.5	2,000	11	15	18

Notes: 1) Values from this table can be customized to fit a particular situation: a. a 2-ton load gives half the application rate of a 4-ton load, all other factors held constant; b. doubling the spacing between passes halves the application rate; and c. doubling the distance of travel halves the application rate. 2) The load size in tons per load can be approximated as 0.03 (spreader volume in cubic feet).

Figure 10.2.   Manure ramp for moving manure from a solid floor to a transport device with minimal labor input

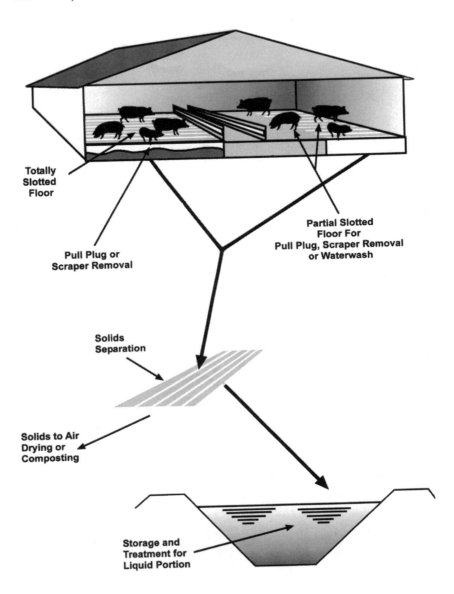

Figure 10.3.   Three alternative manure handling techniques beneath slotted floors. All three techniques have been used successfully. Manure storage beneath slotted floors has proven popular for smaller operations and was the first of the three techniques to gain widespread acceptance.

volume of an underfloor storage pit should be based on the number of animals to be confined, the desired storage period, and an estimate of the daily volume of manure and wastewater to be contributed daily on a per-animal basis. Figure 10.3 shows three alternative underfloor manure handling options. The underfloor storage option is particularly popular for smaller operations and where labor is available for hauling slurry to cropland.

For example, if it were desired to store the manure from 280 finishing pigs, averaging 130 pounds per pig, for 120 days, you might calculate the volume of manure per pig to be 1.3 gallons of urine and fecal material plus an additional 0.3 gallons of cleaning and wasted water. Thus, the storage volume required would be

$$V = (280 \text{ pigs}) (120 \text{ days}) (1.3 + 0.3 \text{ gallons per pig per day})$$
$$= 53,760 \text{ gallons or } 7,200 \text{ cubic feet}$$

Assuming the slotted floor area was 8 square feet per pig, this represents a storage pit capable of holding 3.25 feet of manure between pumping times. In many areas, this is a satisfactory schedule and avoids the need for a separate manure storage facility.

Manure slurries also are stored in pits or basins outside the confinement building. The sizing logic is essentially the same except the period of storage is typically longer and if additional water is going to enter the storage either due to rainfall landing on the basin surface or runoff from an unroofed area, provision must be included to accommodate this additional material. Slurry storage units differ widely in design. There are concrete tanks with slot openings to facilitate scraping manure from a barn. There are earthen basins with an approach ramp so that manure can be pushed into the basin, and there are others in which the slurry is allowed to flow through an open trench. Whatever configuration is selected, it is important that volume be included for all of the material that is added.

There are also aboveground manure storage tanks as illustrated in Figure 10.4. When an aboveground tank is used, the typical arrangement is to scrape the manure from the building into a receiving pit with sufficient capacity to hold 1 day's accumulation, mix the manure into a homogeneous fluid, and pump from the receiving basin into the storage unit. Solids handling pumps, including some open impeller

Figure 10.4.   Aboveground manure storage tank located outside the building and installed when there is a shortage of available space. Aboveground tanks are typically constructed of concrete or coated steel.

centrifugal pumps, are available that can effectively mix the manure and then transfer it into storage. With proper design, it is also possible to use this same pump to agitate the storage unit and pump from the storage unit into the conveyance system to transport the manure to crop or pastureland.

Agitation of manure slurry storages. Storage tanks or basins for slurries should be designed remembering the necessity to agitate the manure prior to its removal. Solids settle during storage of manure slurries and if the material is not homogenized prior to removal, a large volume of solids is left in the storage, making the volume inaccessible to the operator for subsequent storage. Agitation is typically performed by using a chopper pump. A chopper pump chops fibrous material as it enters the impeller. Chopper pumps are essential for beef and dairy wastes where the feeding of fibrous material is common. They also work on other types of manure but it may be possible to use a less expensive open impeller centrifugal pump for swine and poultry wastes in which there is no fibrous or stringy material to create pump-plugging problems.

If the confinement operation is one that minimizes water wastage, it may result in a stored manure that is too dry to be effectively agitated and pumped from the tank. Under these conditions, it is necessary to add water to achieve a slurry that can be moved with the available pumping equipment. If this situation occurs consistently, there are some advantages to adding at least a portion of this dilution water to the storage tank before manure is added. By having at least 6 inches of water in the bottom of the tank, there is less of a tendency for manure solids to stick to the tank bottom and removal is easier. The alternative is to add water at the time the tank is emptied. The latter approach has the advantage of allowing a greater storage period, however, it also has the greatest potential for providing severe mixing problems. Experience indicates which approach works best for a particular operation.

Good agitation requires a pump of sufficient capacity to suspend solids that have accumulated on the tank bottom that may be as much as 25 feet from the discharge point. Pumps capable of delivering up to 1,000 gallons per minute against a head of 100 pounds per square inch (psi) are generally more satisfactory than smaller pumps. The discharge from a pump to agitate a manure storage basin is generally located approximately 6 inches from the floor and oriented to achieve the greatest possible mixing throughout the tank.

The storage unit should be sized to facilitate this agitation. If a pump is to be used for agitation, the tank should not extend more than 25 feet from the pump location. If a larger tank is required, the installation of one or more dividing walls is appropriate so the contents in each portion can be mixed. Smaller tanks, particularly those for swine and poultry waste, also can be agitated by pulling waste from the tank with a vacuum-equipped transport tank and then forcing the material from the transport device back into the storage tank with as much velocity as possible. Alternatively, mechanical agitation with paddle wheels and augers has been successfully accomplished.

Floating scum is typical on storages for dairy cattle wastes particularly if a ration containing roughage is being fed. Many operators have found that this floating scum layer effectively reduces the escape of odors and is therefore a desirable feature. The thickness of the scum layer does not need to be greater than 6 inches to achieve odor control benefits. Thick scum layers tend to be difficult to break up if they dry so they are to be avoided by mixing. Uncontrolled scum layers occupy space that can otherwise be used for storage.

Transport of manure slurries. There are two alternatives for transporting manure slurries from storage to the field where they are to be applied. The more common way is to use a tankwagon or tank truck similar to that

Figure 10.5.   Tankwagons are often used for the transport and distribution of manure slurries onto crop or pastureland. Tankwagons are available with or without soil injection capabilities.

shown in Figure 10.5. Tankwagons are available with capacities ranging from 1,000 to 2,000 gallons. Larger units up to 5,000 gallons are also available but they are limited in their application due to their weight and the resulting soil compaction when they are driven in a field. The alternative to tankwagons is the soil injection equipment in which the slurry is pumped under pressure to a field unit that is towed across the field, injecting slurry as it goes. These pressurized injection devices are relatively inflexible and are useable only when the disposal field is close to the storage unit. A few operators have solved this limitation by having a mobile nurse tank from which the slurry is pumped to the injector, but again, applications are limited.

Perhaps the most common complaint of using tankwagons for distribution of manure to crop or pastureland is the large number of trips required if the operation grows. This is a rational complaint and has prompted many livestock producers to move toward liquid systems in which the manure can be applied with irrigation equipment.

Some tankwagons have knife injectors that are designed to inject the slurry into the soil as it is being distributed. Whether injected from the tankwagon or with a specialized, flexible pipe injection device, the

Table 10.7. Application rates of liquid or slurry manure in inches when applied with a tankwagon moving at 3 miles per hour with a 20-foot distance between successive passes; values in this table may be adjusted for different distances between passes or for different travel speeds

Unloading time in the field, min	Travel distance, ft	Tankwagon capacity, 1,000 gal	Tankwagon capacity, 1,500 gal	Tankwagon capacity, 2,000 gal
0.75	200	0.4	0.6	0.8
1.5	400	0.2	0.3	0.4
3.0	800	0.1	0.15	0.2
5.0	1,200	0.06	0.09	0.12
6.0	1,600	0.05	0.08	0.10
7.5	2,000	0.04	0.06	0.08

advantages are in odor reduction and nitrogen conservation. Where neighbors are a concern, immediate injection may be the option of choice. An alternative is surface spreading with immediate incorporation by plowing or disking the manure into the soil. Neither of these options is as effective as immediate injection, but may be an acceptable compromise when the time for manure spreading is restrictive.

The power requirements to pull a tankwagon across a plowed field at 3 miles per hour range from approximately 24 to 36 horsepower for a 1,000- and 2,000-gallon unit. If a vacuum tank is used and the tank is pressurized to facilitate distribution, an additional 10 horsepower is required. If chisel units are used to inject the slurry 6 inches into the soil, an additional 10 horsepower is needed for each chisel. Thus, pulling a 2,000-gallon vacuum tank across the field, injecting slurry through two chisel units at 3 miles per hour, probably requires approximately 65 horsepower. Power requirements are highly variable depending upon soil conditions so the above-mentioned approximations should be used only as guidelines for planning purposes.

The rate at which manure slurry is applied to a field depends upon the speed at which the tankwagon is moving and the time required to unload. Table 10.7 converts unloading times, travel distance, and tankwagon capacity to manure application rates in inches. Thus, by knowing the nutrient concentration in pounds per acre-inch, you can calculate the appropriate travel distance for unloading a tankwagon. One acre-inch is equivalent to 27,154 gallons.

An alternative to the use of tankwagons for the application of manure slurries is the large-diameter sprinkler, commonly called a gun or big gun sprinkler (Figure 10.6). These sprinklers typically have a nozzle diameter of 0.75 inch or greater and are designed to operate at pressures

Figure 10.6.   Large-diameter sprinklers are able to spread liquid or slurry manure at a much faster rate than can be hauled in tankwagons. Fiber in the manure from the feed or bedding, however, may cause plugging problems. Chopper pumps or screening devices can be used to reduce plugging.

of 70 to 80 psi. Nozzle diameters less than 0.75 inch are prone to plugging when used to apply manure slurries. A typical big gun sprinkler operating at 80 psi would distribute 330 gallons per minute over a circular area with a diameter of 350 feet, an area of 2.2 acres. Under these conditions, this value would represent an application rate of 0.33 inch per hour. Two points become immediately evident. First, the big gun is a rapid way to apply manure slurry to land. One hour of operation would apply an amount of manure equivalent to 20 tankwagon loads at 1,000 gallons per trip. Second, the gun needs to be moved frequently to avoid overapplication to a particular area.

One solution to the problem of needing to move the gun frequently is the use of a traveling gun (Figure 10.7). Conceptually, a traveling gun is a simple gun fitted with a cable and winch system that pulls the gun across the field. Traveling guns are typically designed to be towed either 660 or 1,320 feet. The speed with which they are towed is variable and can be adjusted to achieve the desired application rate. Traveling guns are supplied with slurry by dragging a flexible hose behind them. Units are available in which the flexible hose is stored on a large reel that

Figure 10.7.   A traveling big gun sprinkler is similar to the simpler hand-moved version except that it is mounted on wheels and is pulled through the field by a reel and cable system. It allows spreading the manure over a larger land area without having to manually relocate the sprinkler.

allows the hose to feed out as the gun is moved or to be rolled back on the reel as the gun is towed toward the reel. Alternate irrigation systems are compared in Table 10.8.

One important consideration in using a big gun sprinkler for the application of manure slurries is that the concentration of nitrogen and salts in the slurry is sufficiently high to burn plants. A common practice is to operate the system with clean water for 10 to 15 minutes at the end of each setting. This cleans out the pipe, rinses the outside of that portion of the pipe nearest the sprinkler, and washes the manure solids from the foliage. If the system is not rinsed, a large pool of manure slurry flows out of the pipe when the gun is detached for moving. This pool is likely to be sufficiently concentrated to kill plants that are flooded.

Liquid Manure Handling and Transport

Liquid manure handling is popular particularly for larger livestock operations in which there are economic incentives to substitute mechanical equipment for hired labor. Typical examples are flushed dairy, beef, or swine confinement systems in which the liquid manure is first treated

Table 10.8.   Comparison of alternate irrigation systems for the application of slurry manure to crop and pastureland

Irrigation system	Characteristics	Advantages	Disadvantages
Stationary gun	Single large-diameter nozzle sprinkler	Capable of handling manure slurries	Higher power requirements
	Operates at pressures above 75 psi	Lower initial investment in irrigation equipment	Requires frequent moving, which may be an unpleasant task
		Flexible with respect to land area and application volume	Distribution frequently disturbed by wind
Moving big gun	Similar to the stationary gun except it is towed by a cable toward a winch	Reduces the labor of having to move the sprinkler so frequently	More mechanical parts to maintain, particularly if an auxiliary gasoline engine is used to tow
	Manure supplied by a flexible hose that may be stored on a reel or looped on the ground	Can be used on sloping or irregularly shaped fields	Initial investment greater than the stationary systems

then stored in an anaerobic lagoon or other liquid storage device. Runoff of precipitation from uncovered livestock pens (e.g., feedlots) also generates a liquid waste. Liquid manure under consideration for application to crop and pastureland typically contains less than 2% total solids, is essentially free of solids greater than 0.25 inch in major dimension, and behaves hydraulically essentially as water. There are exceptions, however. Many dairy operators flush free-stall dairy barns and if located in a region where year-round application is acceptable, plan to apply manure to crop or pastureland within 2 to 5 days of flushing. In these operations, solid–liquid separation may or may not precede spreading on land.

Storage of liquid manure. Liquid manure storage is frequently incorporated into the treatment scheme either in a lagoon or a more mechanical treatment sequence. The incorporation of storage capacity into an anaerobic lagoon is discussed in Chapter 7. Runoff retention basins are storage devices to capture and retain runoff due to precipitation falling on unroofed feeding areas. The design of runoff retention facilities is discussed in Chapter 5. Separate liquid manure storage tanks or basins also may be constructed when neither of the other storage alternatives are planned.

Most liquid manure storages are designed so the manure flows into them by gravity from the point of generation. Aboveground storage tanks such as are commonly used for manure slurries are possible but due to the larger volume of wastewater per animal, the cost of pumping generally argues for gravity flow into belowground tanks or basins.

There are typically three components to the volume of a liquid manure storage facility. The first is the process wastewater. Process wastewater is that generated by the normal operation of the facility. This might include clean-up water from a milking parlor, manure being flushed from a confinement building, or discharge from a treatment scheme that is part of the operation preceding the liquid storage. Whatever the source, it is generally possible to determine the daily flow of process wastewater with reasonable accuracy. For example, when designing a storage basin for a flushed confinement facility, it is generally possible to determine the volume of water used in each flush and to determine the number of flushes per day. The volume of manure added can be calculated based on the per-animal production rate, the number of animals served, and the design storage time. The design storage time requires some knowledge of the local area and the frequency with which the operator plans to distribute liquid manure to land. In the extreme case in which once-per-year pumping is planned, the storage interval is 365 days. Storage capacities between 90 and 180 days are more common in an effort to balance convenience against construction costs.

The second factor to consider in the design of a liquid storage facility is the runoff water to be collected during the storage period. If the storage period includes a period of typically heavy rainfall, it may be a significant source of influent, especially if there is an extensive area of unroofed pens contributing runoff. In addition to the normal anticipated runoff during the storage period, good design includes a volume to accommodate the runoff from a design storm for the area involved. Currently, most livestock waste management facilities are required to collect runoff from all storms up to the 25-year, 24-hour storm.

The third component is a volume to hold all precipitation landing on the storage basin. In heavy rainfall areas, this component may be significant because at the time of the heaviest rainfall, the soil is likely to be unsuitable for land application. Table 10.9 summarizes the volumetric considerations in the design of a storage basin for liquid manure.

The details of liquid manure storage facilities are similar to those described in Chapter 7 for runoff retention basins and for anaerobic

Table 10.9. Components of the volumetric capacity of a liquid manure storage basin

Component	Consideration
Process waste flow	This volume represents storage capacity for the wash water, the collected manure, and any other wastewaters that are produced as a normal function of the operation. The usual practice is to establish a daily discharge volume and to multiply it by the duration of storage in days.
Runoff water	If there are any unroofed feeding or holding areas that contribute manure-laden waters to the storage, this volume must be considered in sizing the storage basin. In addition to the normal runoff during the period of storage, it is good practice to include capacity for the design storm, typically the 25-year, 24-hour storm for the area.
Direct precipitation	The storage basin needs to have space to store all precipitation falling directly on its surface during the storage period.

lagoons. The storage should be located with due consideration to having the material enter by gravity rather than by pumping, and it should be located and designed such that extraneous surface water is diverted around rather than being allowed to enter the storage. The bottom of the basin should be well above the upper extreme of the groundwater table. Because the storage basin is likely be in an anaerobic condition, it should be located to avoid being an odor nuisance to neighbors and nearby land users. The inlet can be either above the surface or below, depending upon the needs of the operator.

Like any other artificial surface water body, a storage basin should be constructed with an emergency spillway to protect nearby streams by avoiding an uncontrolled dike failure in case of a storm in excess of the 25-year, 24-hour design storm that could cause overtopping. The emergency spillway should be located to discharge where it will do the least possible damage to area streams and other facilities. A water level measuring post is an assist to the operator in monitoring the volume of accumulated liquid and in ensuring that capacity is maintained for the design storm.

Because the water level in a liquid storage basin is continually changing, erosion of the dikes is of particular concern. Erosion not only robs the basin of storage capacity but also threatens the integrity of the dikes. Some of the problem can be resolved during construction by proper selection of dike slopes and widths and by proper seeding of a densely growing cover crop. In addition, an active program of dike inspection and maintenance is necessary.

Table 10.10.   Advantages and disadvantages of irrigation systems for the application of livestock and poultry wastes to crop and pastureland

Advantages	Disadvantages
Efficient way to apply a large volume of dilute waste to a large area with minimal labor	Initial investment in an irrigation system is likely to be greater than that for hauling a more concentrated waste
Water and nutrients are made available to the crop and can supplement other water and nutrient sources	High level of management is required to avoid overapplication of nutrients and avoid runoff to adjacent land or watercourses
Can be used to apply wastewater of variable quality	Depending upon the waste handling system, the irrigation application system may be an additional source of odor

Application of liquid manure. Liquid manure, whether from an anaerobic lagoon or a liquid manure storage basin, is most frequently applied to crop or pastureland with irrigation equipment. This mode of application allows the large volumes of material to be handled with minimal labor input and to be automated to whatever extent is appropriate for the individual operation. Along with this ability to conveniently apply a large volume of manure with minimal labor comes the management responsibility to apply it in an environmentally responsible manner. Just as the benefits of proper irrigation application are significant, the potential for damage is equally large. Improperly applied liquid manure contaminates groundwater and surface waters, causing fish to be killed. It also impacts neighbor's property, causing severe damage and odor complaints from people living up to 5 miles downwind. All of these problems have been noted in the past so it is important that land application systems be properly designed and operated. The advantages and disadvantages of irrigation disposal systems are summarized in Table 10.10.

The first challenge in the design of a land application system based on irrigation application is selecting an appropriate site. Not all soils and not all locations are suitable. Thus, it is important that in the site analysis the deficiencies of a potential site are identified and that a decision is made as to whether the deficiency can be overcome and at what cost. Thus, site selection becomes an involved process with the need to compromise where compromise is appropriate and to reject sites for which the deficiencies cannot be overcome. Table 10.11 lists some of the factors that are worthy of consideration in the selection of an irrigation disposal site and the constraints inherent in using sites that are less than ideal.

There are several alternatives available for the application of liquid manure to crop or pastureland. The choice of systems is dictated by the land area available, the size of the operation, location, other irrigation

Table 10.11.   Factors to be considered in the selection of a site for irrigation disposal of liquid manures and potential accommodations to make use of a less than ideal site

Factor	Consideration	Accommodation
Proximity to a watercourse or a well used to supply water for human consumption	Runoff from land being irrigated with liquid manure is a significant hazard to adjacent streams and nearby wells; neither fishkills nor contaminated wells are acceptable	Land adjacent to a stream, lake, public water supply, well, or neighbors' pond are not acceptable for an irrigation site; the setback distance required varies depending upon the vulnerability of the resource
Slope of the land	Steeply sloping land presents a greater hazard due to runoff and makes uniform application difficult; low-lying, very flat land may be poorly drained and slow to dry making irrigation difficult or inadvisable	Land with slopes in excess of 10% should be avoided; using an irrigation system with a low rate of application may accommodate less severely sloping land; low-lying, poorly drained soils have limited usefulness for the application of liquid manure
Proximity to neighbors, recreational lands, and public gathering areas	Lands adjacent to neighboring residences, churches, parks, golf courses, commercial areas, or hospitals are at great risk due to spray drift and odor complaints	Appropriate setback distances should be respected in sensitive sites; use of low-pressure distribution systems, immediate incorporation, scheduling according to weather conditions, and a heightened sensitivity to neighbors may allow some sites to be used that would otherwise be off limits
Water intake rate	Area selected should have a sufficiently high water intake rate to facilitate irrigation at reasonable application rates without runoff	Lower application rates can be used but these involve more time for odor release; runoff is an increased hazard with low-intake-rate soils
High-yielding soil	To effectively use nutrients in liquid manure, the soil must have the potential for high yields; highly permeable, excessively drained soils do not allow adequate nitrogen removal nor do they provide the required filtration	Lower and more frequent applications can be made to poor soils; the system is more expensive to operate and presents ongoing concerns
Moderately well-drained soils	To be effectively irrigated, a soil must have a reasonable intake rate to avoid runoff; the drainage rate also must be rapid enough to restore aerobic conditions in the root zone but not so rapid as to preclude nutrient use	Lower intensity irrigation and lower nutrient application rates can be helpful but these limitations may suggest an alternate site

Figure 10.8.   A small-diameter, hand-moved sprinkler is a low-cost option for the distribution of liquid manure. The larger version of this technology is the center pivot irrigation system. There are several small-diameter sprinklers that are moved slowly through the field.

needs of the farm, and the degree to which automation is desired. The simplest option is the hand-moved, small-diameter sprinkler (Figure 10.8). A single hand-moved big gun is the simplest system but probably requires the greatest input of labor and places the most severe constraints on timing of application. A center pivot irrigation system is likely to require less labor and provide a greater degree of application uniformity but represents the greatest initial investment. Each of the systems listed in Table 10.12 has been used successfully. The most important decision is to select a system designed to apply liquid manure at a rate that is acceptable by the soil without surface runoff and that is compatible with the fields and terrain. If irrigation is already in place on the farm, it is appropriate to incorporate manure management into the overall irrigation strategy.

The piping system from the pump to the irrigation site deserves special consideration. Aluminum pipe is commonly used for temporary setups because it can be easily coupled and moved. It also has the advantage of being resistant to sunlight damage. Polyvinyl chloride (PVC) pipe is commonly selected for more permanent buried lines. PVC is sensitive

Table 10.12.   Comparison of alternate irrigation systems for the application of liquid manure to crop and pastureland

Irrigation System	Characteristics	Advantages	Disadvantages
Stationary gun	See Table 10.8	Capable of handling either manure slurries or liquid	Higher power requirements See Table 10.8
Moving big gun	See Table 10.8	See Table 10.8	See Table 10.8
Hand-moved sprinkler system	Series of small sprinklers placed at 40-foot intervals along a lateral Lateral and sprinklers manually moved between sets Moved the distance of the sprinkled diameter	Lower investment than more automated systems Lower pressure required than the big nozzle sprinklers Flexible with respect to field size, slope, and shape Few mechanical parts	High labor requirement of moving individual pipes on a frequent basis Small nozzles may be prone to plugging if solids present
Towline sprinkler system	Similar to the hand-moved system except the couplers are more robust so the lateral may be dragged by a tractor from one setting to another	Similar to hand-moved systems	Requires rectangular-shaped fields due to fixed lateral lengths Driving lanes for the tractor are required in tall row crops
Side-roll sprinkler system	Similar to the hand-moved system except the laterals are wheel mounted and frequently have an auxiliary gasoline engine to move the lateral from one setting to another	Similar to the hand-moved system except avoids some of the labor of moving	Requires a rectangular field that is sufficiently smooth to allow movement Suitable only for a low-growing crop Alignment problems are common

Table 10.12.   *(continued)*

Irrigation system	Characteristics	Advantages	Disadvantages
Center pivot sprinkler systems	Sprinkler laterals are arranged spoke-like from a central hub that slowly rotates Sprinklers are spaced to achieve uniform distribution Various nozzle designs are available to customize application and droplet size	High degree of automation therefore minimal labor required High degree of uniformity in application rates Come in sizes to accommodate up to 640 acres	High initial investment Sophisticated mechanical system that requires high-quality management
Surface irrigation, controlled flooding	Small earthen dikes are constructed around level fields or portions of fields; manure is allowed to flow into the area and be distributed throughout the area	Very little equipment required	Difficult to achieve uniform distributions of water and nutrients Requires level fields and special dike construction
Surface irrigation, furrow	Furrows are formed in the field generally along contours; liquid manure is distributed to the furrows with gated pipe such that the pipe openings match the furrows	Very flexible system for small operations expecting to enlarge Low Power requirements	Intensive management is required to achieve uniform distribution, and avoid excessive nutrient application in certain areas

to ultraviolet and should not be used in aboveground applications. It has a smooth interior surface so there is less headloss than in aluminum pipe. Polyethylene pipe is available in small diameters and is a low-cost alternative for small systems. It has the disadvantage of using interior couplings so is more prone to plugging.

Whatever the piping material selected, it is appropriate to consider how the system will be protected from freezing damage and from high-pressure damage if it should plug at some downstream location. A pressure relief valve should be installed on the discharge side of the pump and be set slightly above the normal operating pressure. Air relief valves should be installed on underground PVC pipelines at every high point. Two conflicting challenges face the designer in planning the irrigation piping system, particularly if it is one of considerable length. The first is the necessity to maintain a sufficient velocity in the pipeline to prevent solids settling. Minimum velocities of 5 to 6 feet per second are suggested. Second is the matter of not having so much headloss that the irrigation sprinkler operation is adversely impacted. The activation of a high-pressure irrigation system deserves special consideration. Electrically driven pumps in particular reach full delivery very quickly. Hitting the sprinkler heads with the full load in a short interval has created problems for some operators. This problem can be avoided by the installation of a butterfly valve on the discharge side of the pump or by otherwise slowing the rate at which the line is brought to full pressure.

The pump and power unit should be selected to be appropriate both to the material being pumped and to the setting in which it will operate. Most liquid manure irrigation applications use a simple open impeller centrifugal pump because of its high efficiency and flexibility of operation. If a more concentrated material is to be pumped, alternate pumps particularly designed to handle solids should be considered. Helical screw, rotor-lobe, and piston pumps have all been used to pump manure slurries with some degree of success. Each of these pumps is a positive displacement pump and develops high pressures, if necessary, to maintain the flow. These pumps are all considerably more expensive than centrifugal pumps and easily damaged by solid objects that may have entered the manure such as tools, bolts, or stones.

11

Alternatives to Manure Application to Crop and Pastureland

Manure has traditionally been applied to crop and pastureland in the United States and western Europe as a source of plant nutrients, as a means to maintain or restore soil tilth, and as an environmentally acceptable final disposal mechanism. This paradigm has dominated most western research efforts devoted to manure treatment and management. There are alternative systems in operation and as livestock and poultry production becomes more regionally concentrated, these alternatives will be given greater consideration.

In much of Asia where land holdings are small and population densities are high, confinement livestock production has developed on the premise that the effluent after treatment can be discharged into surface watercourses flowing toward the ocean. There are both corporate and individual family farms in this system and the confinement facilities cover the greater part of the land holding. Livestock producers are not crop farmers nor is there a demand for the liquid manure from systems on adjacent or nearby cropland. As a result, livestock and poultry producers are designing and building livestock waste management systems that focus on other products rather than a soil-amending manure. The typical choice is a highly treated liquid effluent that meets existing water quality standards and a dry, composted material made from the solids that have been processed to uniquely meet an economic demand among the farming or nonfarming community.

The U.S. alternative is the large, integrated livestock producer of the upper Midwest raising more than 100,000 pigs in a confinement system in which the manure is flushed from the building two to three times per day. The treated liquid effluent is reused as flushing water, whereas the solid fraction is composted and marketed to the ornamental nursery industry as an ingredient to standard potting soil mix. In this system, the incentive is to minimize the required land. Frequently, these operations have developed agreements with neighboring landowners to apply manure or treated liquid manure to the neighbors' land. This arrangement has worked well in some instances but is limited to landowners in proximity to the livestock operation.

Another U.S. alternative is a small-scale livestock operation for which owning specialized irrigation equipment is not compatible with either the scale of the operation or the surrounding land area. A combination of wastewater treatment with solids separation and a medium filter or overland flow to economically produce nitrate followed by an engineered wetland provides an environmentally acceptable option for removing nitrogen by denitrification and thus reduces land required for nutrient use. For these and the other nonconventional manure handling systems, the important question is, What are the alternative uses for manure that are environmentally acceptable and for which the cost of processing is compatible with the economics of competitive livestock production?

Refeeding of Manure to Other Livestock and Poultry

The direct refeeding of poultry manure both to poultry and to ruminants has been studied in the research community and shown to be effective in demonstration-scale operations, however, the adoption of this practice has been limited.

Either a visual or a chemical examination of animal manure or poultry droppings is sufficient to confirm that an animal has not gleaned all of the potential nutrient value from the feed and that it would have value as a feed ingredient. Poultry were raised in the past by allowing the chickens to follow the cattle and to pick through the manure selecting their ration. Satisfactory poultry production resulted with a minimum of purchased feed. Thus, there is nothing inherent in the manure that makes it unsuitable for incorporation into an animal ration.

Process

Among the earliest research efforts on manure refeeding were studies on the refeeding of dehydrated cage layer droppings. The droppings were mechanically collected by scraping from a dropping board and fed into a gas-fired dehydrator, similar to that used to dry hay pellets. The dryer was operated at slightly more than 140°F to ensure that potential pathogens were killed. This material was fed to chickens in a study involving several generations of birds to ensure there was no long-term accumulation of growth retardants or toxic materials. Because of the lower metabolizable energy content of dried poultry droppings compared with standard feed rations, it was not effective to use dried manure as the only feed ingredient, but it could be successfully incorporated at a level of 15 to 20% of the total ration.

Several sources of manure, including poultry droppings and beef cattle wastes, have been mixed with hay, corn fodder, and corn grain to produce a silage-like material commonly called wastelage. A typical process would include collecting beef cattle waste either daily or every second day and mixing this fresh manure with corn and corn fodder and subjecting this mixture to 10 or more days of ensiling in an oxygen-free environment. A mixture of up to 40% manure can be successfully ensiled. The material removed from the silo has the appearance, smell, and consistency of conventional corn silage and is readily consumed by cattle. The final product, or wastelage, has been successfully fed to cattle. Thus, cattle consume the product and the incorporation of either poultry droppings or cattle manure effectively provides nutrients to the feed that are being used by the cattle. But the adoption of the process has been limited.

Other direct refeeding options have been considered but generally found less satisfactory. Refeeding of the solid fraction from a solid–liquid separator was studied but it was confirmed that the most desired nutrients for refeeding were contained in the liquid fraction. Hence, there was little benefit in refeeding the low-energy material claimed from the separator. Composted manure mixtures are similarly too low in energy value to make them valuable as feed ingredients.

Constraints

With the successes that have been achieved in the direct refeeding of manure either to the same species or to an alternate species, the question must be asked as to why there has been so little adoption of the process in U.S. commercial livestock production. There would seem to be several partial answers to this question but none that have been judged sufficiently compelling to eliminate the concept from continuing consideration should economic and cultural parameters change.

First, there is the aesthetic consideration. Meat, poultry, eggs, and milk products are sold in the United States as high-quality products suitable for the most demanding consumer. There is a concern that if there was widespread manure refeeding, the products would be considered less suitable for the most discerning consumer. This reaction is in spite of all the research results that confirm that manure feeding does not alter the flavor or nutrient value of the flesh or other animal products.

Second, processing of manure is an additional cost-creating step in the animal production process and to be attractive to a modern livestock producer, must result in the production of a product of the same or better quality at a lower cost. Current manure refeeding processes are neither convenient nor is there manufactured equipment readily available to facilitate the refeeding process. Feed ingredients are readily available in the United States at costs that allow meat and animal products to be marketed at prices the consumer is willing to pay. Processed manure would need to be available at a lower cost than the ingredient it replaces to be competitive. Currently, the cost incentive for the use of processed manure as a feed ingredient is not sufficient to promote its development.

There are also some regulatory constraints that currently restrict the marketing of processed manure in interstate trade because of its identification as a feed additive. Although progress has been made in easing this restriction for dried poultry manure, there would need to be additional quality assurance efforts to gain approval for unrestricted sale and distribution of manure-based products as feed ingredients for either meat-, milk-, or egg-producing animals if these products are destined for the commercial market.

And third, refeeding of livestock and poultry manure might reduce the cost of waste management or be used as a way to lessen the environmental impact of concentrated livestock production. These benefits have not been demonstrated. There has been little in the refeeding results to suggest that one or more of the conventional waste management steps would be eliminated or the cost significantly reduced. Ensiling manure has the potential to reduce hauling costs; however, when the process is investigated further, current silage making is a batchwise process and is done as a way to store the corn or hay crop when it is available. Thus, corn silage is normally made in an early autumn time frame of 2 to 4 weeks. Manure, in contrast, is available on a continuous basis and if stored loses much of its value as a silage ingredient. Hence, there is no significant reduction in overall waste management expenditures by incorporating manure in compost for a few weeks of the year.

Conversion of Manure Constituents to Alternate Feedstuffs

Based upon the chemical composition of livestock and poultry manure, one of the most logical steps would be to convert one or more of these constituents into an alternate but higher value product. Several alternatives have been investigated and many of them are being applied in specific situations where the economics are particularly attractive or where other factors dictate against traditional practices.

Aerobic treatment of animal manure produces an aerobic bacterial cell mass that is high in protein. This material has been fed to pigs as a liquid and favorable results have been reported. The major limitation of the system was the reported dilute concentration, hence limited intake.

Fish ponds fertilized with animal manure are common in much of the world. The solid particles can be consumed immediately by the fish.

Nutrients support the production of algal cells that are used as food by the fish. This system is suitable only when a limited number of animals is involved, otherwise maintaining an adequate dissolved oxygen concentration in the pond becomes difficult. Under this system, the waste from 10 pigs is sufficient to fertilize a pond of 1.7 acres. Annual fish yields of up to 400 pounds per acre have been reported. Thus, this system would yield up to 20 pigs per year and an additional 700 pounds of fish. Higher fish yields are possible if raceway or pond production units are aerated to supply additional oxygen.

The effluent from an anaerobic digester can be fed to shallow, stirred algal growth basins for nutrient recovery. Large and sustained algal production was achieved with this type of system. Such a system was operated on a sustained basis in Singapore during the 1980s. Algal production rates in excess of 275 tons of dry matter per acre per year have been reported. The difficulty with this system has been the high cost of algal recovery and drying.

In a similar system, water hyacinths have been grown on anaerobic digester or anaerobic lagoon effluent, producing a large annual yield. Although easy to harvest, the water hyacinths proved difficult to dry and were not found suitable for direct feeding due to their high water content and low density. They were effective, however, in recovering a large fraction of the nitrogen being fed into the water hyacinth ponds. Some interest has been expressed in the harvesting of water hyacinths and subsequent anaerobic digestion of the plant tissue to produce biogas. This concept has not seen commercial development due in part to the relatively low cost and availability of more convenient fossil fuels.

From these examples, it is clear that dilute or partially treated animal manure is suitable for several alternate biomass production schemes. Some of these schemes are in use in specific locations where local conditions make them economically attractive. None has, however, achieved widespread application either due to the low economic value of the end product or due to the extensive resources necessary to construct and operate the processing facility.

Overland Flow Alternatives

Although they have some similarities to land application, there are two alternative overland flow treatment and disposal options that have been used. They have application in situations where conventional application might prove impractical or unreasonably difficult. The first alternative, commonly called the grassed waterway, grass filter, or infiltration terrace is particularly appropriate for livestock operations with a

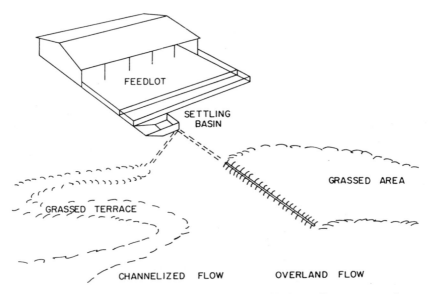

Figure 11.1.   Grassed waterways have proven to be suitable for small waste sources for which more highly engineered systems would be inappropriate.

small number of animals or for runoff from an unroofed feeding area in which the amount of collected runoff does not justify investment in an irrigation system. The second overland flow alternative is indeed a treatment option and the overflow from the treatment is collected and subjected to conventional land application. The benefit of the overland flow treatment is to reduce the nutrient and organic loading sufficiently to allow a smaller land application area than would otherwise be required.

Grassed Waterways

Grassed waterways have been demonstrated as an effective alternative for disposal of a relatively small volume of unroofed feeding pen runoff or dilute liquid manure when the volume is less than 1,000 gallons per week. Figure 11.1 shows a typical application in which the runoff from a feedlot of less than 1 acre is drained into a settling basin which in turn discharges into a serpentine terrace. The terrace is designed to exclude other runoff and to have a sufficiently low velocity so that the wastewater can be absorbed during the flow process. Table 11.1 provides a guide in selecting the length of terrace based on the slope. Grassed waterways are typically built with flat bottoms and sloping sides, resulting in trapezoidal

Table 11.1. Minimum lengths for grassed waterways to provide for the infiltration of feedlot runoff based on the slope of the grassed waterway

Slope, %	Flow length, ft
0.5	300
0.75	375
1.0	430
1.25	475
1.5	525
2.0	625
2.5	750

Table 11.2. Anticipated performance of a grassed filter system treating feedlot runoff (values are percentage of reduction in both concentration and mass of the constituents measured)

Constituent	% Reduction, concn.	% Reduction, mass
Ammonia N	85	98
Total Kjeldahl N	80	97
Total solids	70	96
Chemical oxygen demand	85	98
Phosphorus	78	96
Potassium	75	96
Fecal coliforms	95	97
Fecal streptococci	95	97

cross section. The width of the terrace bottom is selected to provide the area necessary for infiltration of the runoff plus direct rainfall on the waterway.

The concept of a grassed waterway or similar design is to create an area of dense vegetation in which the liquid manure or manure-laden runoff can flow at low velocities, be treated by the vegetation, and the volume be reduced by infiltration and evapotranspiration. It is important that provisions be included for the maintenance of the grass cover and for the distribution of the contaminated water so the entire area allocated to the treatment process is used. Depending upon the soils involved and the rainfall pattern, filter areas typically have 0.5 to twice the area of the feedlot from which they receive runoff. Because filter areas operate intermittently and are dynamic, typical performance evaluation is difficult. The data in Table 11.2 are provided as an indication of what might be expected.

Consideration of the grassed waterway as a treatment device is instructive for the potential design of specific pollution abatement measures. Additionally, the results reported in Table 11.2 support the widely held perception that vegetated filter strips along the lower edges of manure application areas have significant value as additional protection against water quality damage from a spill or a rainfall event during or immediately after land application. Prudent design frequently includes a grassed buffer strip of at least 50 feet between manure application areas and waterways as an additional level of protection. This is consistent with the experience in grazing areas that indicator bacteria, fecal coliform, and fecal streptococcus seldom travel more than 6 feet across the surface of a well-established pasture or other grazing area.

Overland Flow Treatment

Overland flow treatment has some similarities to grassed waterways but the intent and the design are drastically different. In an overland flow treatment scheme, the goal is usually to reduce pollutant concentrations, particularly the total nitrogen concentration, to allow it to be safely applied to a land area that is insufficient to otherwise accept the manure load. Thus, by applying the liquid waste to a land area and allowing it to pass over the surface and contact the vegetation on the land, treatment occurs and the nitrogen, phosphorus, organic matter, and microorganism concentrations are reduced. At least half and frequently 80% of the applied wastewater flows off the lower end of the overland flow treatment system and needs to be collected and applied to crop or pastureland as the final disposal procedure. In addition, any rainfall running off the overland flow site requires capture and land application. A typical overland flow treatment system is shown schematically in Figure 11.2.

Overland flow schemes are particularly attractive when land areas available for liquid manure application are sufficiently impermeable to restrict application rates. They are also attractive where the availability of land with adequate permeability and productivity is limited and a less permeable area can be used to lower the concentrations to allow safe disposal on the limited amount of more desirable land.

Overland flow treatment is applicable to dilute manure slurries such as might be collected from a feedlot runoff retention basin, lagoons, or a flush system in which the liquid for which disposal is being sought contains less than 2% total solids. Soils of low permeability, less than 0.3 inches per hour, have generally been selected for use in overland flow systems to avoid downward transport of excess nitrogen into the groundwater zone as might occur if a more permeable site were to be used in this area. Clay or fabric liners may be added to reduce the infiltration rate where necessary.

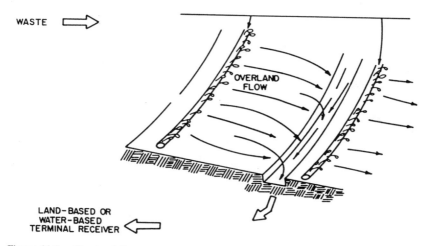

Figure 11.2. Overland flow treatment systems are designed to make use of the biological and physical treatment capabilities of a vegetation-covered soil surface. Infiltration is not a major objective.

Overland flow process. The overland flow treatment process involves several of the traditional mechanisms already discussed in this book. Because of the vegetative cover on the overland flow site, the liquid layer flowing across the soil surface is subjected to numerous settling opportunities and there is a physical filtering due to having the thin layer of liquid pass across the rough soil surface. Thus, it is not unexpected that there is a net reduction of particulates. This reduction is enhanced if the design ensures that the flow velocities are sufficiently low to avoid erosion. The thin layer of liquid flowing across the vegetated soil surface creates a microenvironment similar to that on the surface of trickling filter stones where a bacterial population feeds on the organic matter in the wastewater. The thin layer allows oxygen to permeate the liquid, creating in effect, a fixed film aerobic treatment site, thus explaining the observed decrease in the dissolved organic constituents.

Nitrogen concentrations are reduced in an overland flow treatment process by several mechanisms. Organic nitrogen is associated with the solid particles. Thus, solids settling and filtration affect organic nitrogen reduction. The large surface area promotes the volatilization of ammonia and the conversion of the ammonia fraction not volatilized into nitrates by biological oxidation. Along with the nitrification, there are likely to be sites of sufficiently low oxygen availability to promote biological denitrification, converting a fraction of the nitrates formed into nitrogen gas that volatilizes to the local environment.

Phosphorus, typically associated with particulate matter in the liquid manure, is trapped by the settling and filtration processes and bound to the soil particles near the soil surface. As long as the system is designed and operated to minimize soil erosion, phosphorus removal should be in excess of 60%. Phosphorus removal is reduced over time as the ion exchange capacity of the surface soil is exhausted.

Indicator bacterial concentrations are reduced both by the settling and filtration processes and by exposure to sunlight. Research in other settings supports the concept that thin sheet flow of liquid manure across a soil surface achieves fecal coliform and fecal streptococcus removals in excess of 95% in less than 25 feet.

Design of overland flow treatment. Overland flow systems have been constructed to serve a variety of municipal, industrial, and agricultural applications. The systems have been nearly as diverse as the applications. Common to the systems has been the distribution of the wastewater along the upper edge of the treatment site at a flow rate sufficient to produce a flow at the lower edge collection channel of at least half the application rate but at a sufficiently low rate to avoid erosive velocities. Overland flow sites have been successfully operated on sites sloping from 0.5 to 5%. Slopes of 1 to 2% are probably easier to manage. The length of flow is typically greater than 50 and less than 400 feet. Flow lengths of 100 to 200 feet are probably most appropriate. Overall application rates vary from less than 0.4 to more than 4 inches per week. Application rates are probably best based on experience with local soil conditions. At least half the applied volume should be collected at the lower end of the site and there is a need to have velocities sufficiently low so as to avoid erosion within the treatment area. Application of wastewater to an overland flow site is always an intermittent operation. Intensively used sites might have liquid applied twice a week. Agricultural operators might well decide to apply wastewater only during the growing or nongrowing season and during this period, twice monthly would be the preferred frequency.

Overland flow treatment of liquid animal manure has been established as an option to reduce nutrient, organic matter, and indicator bacterial loads but does not typically produce an effluent suitable for discharge to receiving streams. The extent of treatment can be influenced by the design. Lengths of 100 to 400 feet are typical, lengths of up to 1,000 feet have been used. Table 11.3 provides some guidance as to the range of treatment to be anticipated. The table shows a range of values. The higher pollutant removal rates are generally associated with longer overland flow systems and lower application rates.

Generally, overland flow systems, because of their intermittent loading and their low application rates, can be operated without the creation

Table 11.3. Anticipated ranges of pollutant reductions for overland flow treatment systems (based upon treating a dilute liquid waste similar to that from a flushed manure system after solid–liquid separation or storage and treatment in an anaerobic lagoon)

Constituent	Range of removal, %	Anticipated effluent concentration, mg/l
Water volume	40–60	
BOD	60–80	100–300
Total suspended solids	60–80	150–500
Total nitrogen	40–80	200–500
Ammonia nitrogen	70–90	20–100
Total phosphorus	40–80	5–20
Fecal coliform	85–97	0.01–1×10^6/100 ml

of highly objectionably odors. It would be appropriate to consider this factor in design, however, because large amounts of ammonia can be volatilized and a large surface area created.

The advantages of an overland flow system are that it can reduce the amount of land required for manure disposal and may, in certain situations, produce an effluent of sufficiently high quality for reuse as a flush-water. Disadvantages are that it adds an additional processing step and may further contribute to odors in the area.

Wetland Treatment

Within the past 20 years, the importance of natural wetlands in the preservation of water quality has become evident and the earlier widespread practice of filling and draining wetlands has been curtailed. Wetlands are currently protected in the United States by a series of state and federal regulations. Wetlands and swamps collect runoff and shallow groundwater and provide long-term storage in the presence of a variety of plant species that grow in waterlogged or submerged soils. During this process, the water moves slowly toward surface streams, providing much of the dry weather flow. During this period of contact with the plants, nutrients are captured, particulates are settled, and even more resistant organic molecules degraded over a period of time. In areas where wetlands have been filled or drained for either crop production or economic development, lower, sustained dry weather flows have been noted along with a decrease in water quality.

Engineered wetlands have been constructed in many areas of the United States to simulate this natural treatment process as the final treatment step for a number of wastewaters (Figure 11.3). Partially treated

Figure 11.3.   Constructed wetlands have been incorporated in livestock waste management systems where alternate land application processes proved inappropriate and direct discharge to surface waters was unacceptable.

sewage or waste stabilization pond effluent, still containing dissolved nitrogen and phosphorus in sufficient concentrations to contribute to excessive algal growth in downstream impoundments, has been subjected to this process. Nutrient concentrations, indicator bacterial populations, and more resistant organic species concentrations were reduced, particularly during the warm growing season. In many locations, this was also the season of greatest water quality concern so the process has been credited with having improved a number of perplexing water quality concerns. For several small communities, engineered wetlands are proving to be relatively low-cost, low-energy, and minimal-maintenance alternatives to tertiary treatment options.

The design of constructed or engineered wetlands is continuing to evolve as the applications become more diverse and the treatment objectives better defined. Two basic approaches are being used. One is the open water wetland, the other a submerged wetland.

In the open water wetland, the area of the wetland is constructed so the water pools above the soil surface to a depth of 3 to 12 inches. Natural or specially selected aquatic plants that grow with their root systems submerged, such as common cattails, bulrushes, and a variety of wet-soil grasses are established during the construction of the wetland. Once wastewater is introduced, natural selection tends to move toward a mixed plant population appropriate to the area and the environment created within the wetland. These constructed wetlands are generally designed to have a hydraulic retention time between 30 and 90 days.

Submerged wetlands are designed similarly except that the area intended for the water is filled with a porous material such as sand and gravel to provide a support structure at the surface for plant root systems and to eliminate the free water surface. The submerged wetland concept has some advantage in areas were mosquito control is important. The disadvantage is that more land area is required to achieve the same hydraulic retention time.

Performance data on constructed wetlands is difficult to transfer from one location to another because of the variabilities in growing season, plant populations, and waste differences. The overall conclusion, however, supports the use of the constructed wetland concept in locations where other alternatives provide an inadequate level of treatment, or where suitable lands are available. In the more successful situations, effluent is discharged from the wetland only during precipitation and the quality of the effluent is comparable with natural wetland overflow, low in nutrient concentrations, adequately aerated, and essentially free of indicator enteric organisms.

Application of Constructed Wetlands to Animal Feeding Operations

Constructed wetlands have been incorporated in several animal feeding operations around the United States as a means of reducing the nitrogen and phosphorous concentrations to allow application to a smaller land area. Table 11. 4 presents a summary of data collected from 68 sites. Of these sites, 46 were dairy and cattle feeding operations, 19 were swine operations, and the other 3 were poultry and aquacultural operations. In each case, relatively dilute wastes were directed to the wetland treatment. For the cattle operations, the wastes tended to be surface runoff or diluted flushwater. The swine wastes were typically from an anaerobic pretreatment lagoon. As Table 11.4 indicates, the constructed wetlands effectively reduced the nitrogen and phosphorus concentrations, thereby reducing the land area required for disposal.

Pretreatment of wastes prior to wetland treatment. Concentrated wastes need to be pretreated prior to introduction to a constructed wetland to eliminate solids accumulation in the inlet area and to ensure that high concentrations of organic matter do not adversely impact the wetland ecosystem. Dilution water also may be required if the concentrations of nitrogen or organic matter are high or if there is a high rate of evaporation from the system. Several pretreatment options have been used.

Table 11.4. Average inflow and outflow concentrations as well as reduction for 68 different constructed wetland systems used to treat wastewaters from confined animal feeding operations

Wastewater constituent	Average inflow concentration, mg/l	Average outflow concentration, mg/l	Average % reduction
Biochemical oxygen demand, BOD_5	263	93	65
Total suspended solids, TSS	585	273	53
Ammonia nitrogen, NH_3-N	122	64	48
Total nitrogen, TN	254	148	42
Total phosphorus, TP	24	14	42

- Lagoons are used to settle solids and allow phosphorus to precipitate and settle. Lagoons also reduce nitrogen concentrations, allow organic matter to degrade, and temporarily store wastes for flow equalization.
- Storage ponds, generally smaller than lagoons, can be used either as an alternative to lagoons or in addition to lagoons to accomplish many of the same functions but to a lesser extent.
- Solid–liquid separation devices also have been used to prevent solids accumulation in the wetland. Solid–liquid separation devices typically remove 40 to 60% of the solids and a lesser fraction of the nutrients.

Design of a wetland treatment device. Wetland loading rates vary considerably depending upon local climatic conditions, type of incoming wastes, and desired effluent quality. Typical nitrogen loading rates are between 5 and 25 kilograms per hectare per day. Organic loading rates are based on biochemical oxygen demand and typically range from 40 to 200 kilograms per hectare per day. Retention times are generally from 4 to 15 days.

Wetland cells can be shaped to meet the demands of the local terrain. Treatment results are best when flow is evenly distributed across the cell. Length-to-width ratios range from 1:1 to 10:1.

Embankments must be constructed using appropriate design and compaction techniques. Embankments should be 1.5 to 2 feet above the highest expected water level to contain severe rain and to accommodate sediment accumulation. Dikes should have side slopes that can be maintained with tractor-mounted mowers. Bottoms of constructed wetlands should consist of a compacted clay layer or a liner covered by 12 inches of topsoil to serve as a rooting bed for plants.

Water depths are typically 1 foot or less in constructed wetlands. It is frequently wise to include flow control devices that permit adjusting the water level between 0.1 and 1.5 feet.

Most wetlands used for the treatment of animal wastes are constructed open water wetlands. Natural wetlands are not appropriate for the treatment of animal waste.

Reciprocating Constructed Wetlands

Reciprocating wetlands refers to the recurrent filling and draining of adjacent wetland cells, thereby promoting development of alternating anaerobic and aerobic zones. With reciprocating wetlands, it is possible to manipulate the treatment systems biological structure and function when nitrification and denitrification are sought to facilitate nitrogen removal to minimize land application area requirements. Reciprocating wetlands are expensive to construct and require additional management attention to achieve optimal performance, but they provide the manager with a variety of options to achieve results that could not be obtained with conventional constructed wetland treatment.

Phytoremediation Systems

Trees like other plants have an ability to use large quantities of water and to effectively capture nutrients. Application of animal waste to trees has the advantage of an infrequent harvesting schedule and the presence of a deep root system. Trees also may be grown on land that is not suitable for more-conventional row crop production.

Recent interest has been in the planting of rapid-growing hybrid poplar trees that reach harvest size in 5 to 7 years. Poplar fibers are used in the manufacture of paper.

In a typical phytoremediation plantation, poplar saplings are planted in rows approximately 10 feet apart. The spacing between trees may only be 1 to 2 feet. Poplars have several advantages in addition to their rapid growth. One of these advantages is that a poplar can be started by placing switches in a hole or trench and allowing the switch to develop roots and to grow. A well-managed poplar plantation area is capable of extracting more water and nitrogen from an area than a conventional crop and the deep roots provide additional assurance that nitrate cannot escape to the groundwater. Poplars are most often harvested by clipping just above ground level. The wood is then used as a source of fiber in the manufacture of paper products. Thus, the application to livestock waste management is most attractive in areas where there is ready access to an established pulp processing facility.

In addition to the use of phytoremediation as a disposal and treatment option, several installations of poplar trees have been selected for extracting contaminants from landfills and similar buried waste materials as a way to prevent the escape of leachate. Plantations also have been installed between abandoned animal feeding areas and nearby streams to intercept the flow of nitrate-bearing shallow groundwater that was otherwise flowing into the stream as the major component of base flow. Nitrogen extraction rates of more than 1,000 pounds per acre per year have been observed.

Biofilters

The passage of partially treated animal waste through composted organic materials is an additional option for removing the final traces of organic matter, nutrients, and more resistant residuals. Although the design of such facilities has yet to become fully established, applications are arising in which effluent from more conventional anaerobic and aerobic treatment processes are allowed to flow slowly through these organic beds to further reduce odor emissions and enhance quality before allowing the material to discharge to the environment. This process is similar to the slow sand filters that have been used in early municipal wastewater systems, except that the presence of composted biomass provides a rich and varied source of microorganisms and an energy source to microbial species capable of feeding on the color-producing compounds and other stable organic molecules.

Physical–Chemical Conversion to Fuel

There are physical–chemical processes available for the conversion of the organic components of animal and poultry manure into fuel products. In general these processes involve subjecting a "cleaned" and dry manure to high-temperature, high-pressure treatment to chemically reorganize the molecules into hydrocarbon-resembling species that can then be separated in manners similar to those used in the conversion of crude oil into petroleum products. Investigation of these possibilities was encouraged during the 1970s when it appeared that alternate energy sources were to become a priority. With the increased availability and relatively low cost of oil from foreign sources since then, there has been little incentive to pursue these options.

Pyrolysis is a high-temperature chemical process in which raw organic material, such as animal manure, chemically decomposes under temperatures ranging from 400 to 1,500°F. The resulting material can be separated based on melting and boiling points to produce a liquid fuel, a char, and an aqueous phase. When conducted at atmospheric pressures and a temperature of approximately 1,500°F, the pyrolysis products resemble synthesis gas that is primarily a mixture of carbon monoxide and hydrogen.

12

Odor Control

The development of large concentrated livestock and poultry facilities has led to a marked increase in concern about odors that may be discharged to the environment. This situation has been further magnified by a trend toward the development of nonfarm residences in previously rural areas. Thus, there has been an increase in the size of animal production units while at the same time there has been a growing expectation of an odor-free environment among rural residents. This has led to a demand for information and technologies to reduce odor emissions. This demand is somewhat different than the demand for water pollution control because of the difficulty of defining an objectively measurable, acceptable level of odor release.

Research has demonstrated odor has at least two aspects. One is objective and measurable both in concentration and duration. The other is subjective and related to offensiveness. How good or bad an odor is judged to be is related to individual preference and previous experience. Observations over the past 20 years suggest that people who are less acquainted with livestock odors, who have another interest in not smelling the odor, or who have some other basis for negative feelings toward a particular livestock operation, are generally more likely to complain of odors. People who see confinement livestock and poultry production as a contributing component of the local economy are generally more tolerant.

Technologies exist to produce livestock and poultry with a degree of odor control that is judged to be acceptable. The challenge to the designer of facilities is related to the cost of systems that provide a high degree of odor control compared with systems that do not. The situation is further complicated in that not all locations require the same degree of odor control nor are the requirements stable over time. Larger operations have a greater odor production potential and the technologies appropriate for small operations may not be acceptable for their use.

Odor control challenges and appropriate reduction technologies differ for the various modes of production. For example, odor control measures for an unroofed beef cattle or dairy feeding area located in a

257

relatively arid area are dramatically different than those for a roofed confinement facility with a flushed manure transport system. Both of these systems present different problems than a poultry feeding operation in which pullets are raised to broiler weight on a litter-covered floor. Each of these and the many other systems in use require special and individual odor control consideration.

Unroofed feedlots may have odors that arise from the storage of putrescible feed materials. This is a particular concern for cattle feeders using food processing wastes as ration ingredients. The more widespread odor source from feedlots is that from manure decomposing on the feeding surface. Moisture control becomes an important consideration in resolving both feedlot surface odors and dusts that may be a similar nuisance if the surface becomes excessively dry. Another odor source associated with unroofed feedlots is the runoff collection, storage, and disposal system. The odor from stored feedlot runoff is frequently similar to that from an anaerobic lagoon and when this accumulated liquid manure is field-applied, typically with conventional irrigation equipment, there is considerable odor release. The land application odor is an infrequent event but that from the retention basin may be more nearly continuous.

Two manure management schemes for pork production currently predominate in the United States. One is the storage of manure as a slurry in one or more pits or tanks immediately beneath the confinement facility or in a tank or pond located outside the facility. Poultry also may be fed in buildings in which the droppings are handled as a slurry. Dairy manure also is frequently handled in similar systems. The goal is to transfer the manure into the storage facility with a minimal dilution with water. There may be a continual release of odors from the storage unit and there is a considerably more intense release of odors at the time of agitation in preparation for field application. One odor control alternative is to cover the manure in storage. The cover may be an impermeable cover such as that involved in an anaerobic digester for the production of biogas or a permeable cover designed to biologically oxidize the odorous gases as they pass through the covering material.

A second popular manure management system for pigs and poultry includes a liquid handling system in which the manure or droppings are treated and stored in an anaerobic lagoon. Lagoons provide a low-cost means of manure storage and result in a large portion of the nitrogen originally in the manure or droppings escaping to the overlying air. Proper lagoon design and management are intended to achieve a relatively low intensity of odor release. This approach is effective much of the time, however, during late spring and early summer as water temperatures increase, elevated odor levels are more frequent. As lagoons become larger, the odor-releasing surface becomes sufficiently large to present an odor problem to sensitive surrounding landowners at increasing distances from the facility.

An alternative manure management approach based on having the manure dry is popular in the confinement of poultry. Broilers are typically raised on bedding to absorb the droppings. The heating ventilation system is designed to provide sufficient drying to maintain the moisture content low enough to preserve aerobic conditions. With water leaks or other equipment malfunctions, wet areas and ensuing anaerobic conditions may develop. Caged poultry houses, typically used for layers, may store droppings in a pit beneath the cages that is designed to be sufficiently dry to avoid anaerobic odor production. Leaking water systems or inadequate heating and ventilation can cause wet areas and elevated odors in these facilities. The dry manure systems, whether litter-based or deep dry storage, result in the release of elevated odors at cleaning time, which may be one to three times yearly.

As this discussion indicates, most livestock and poultry operations operate without encountering odors in sufficient concentrations or with sufficient frequency to cause complaints from neighbors. Neighbors may not consider the odors sufficiently strong to interfere with their enjoyment of their property or the odors may be perceived as a "normal and expected" aspect of living in the area. Although the fraction of livestock producers who encounter odor problems is relatively small, the expense of dealing with these odors is sufficiently great that most designers acknowledge the importance of incorporating an odor control strategy in the initial facility design.

Odor Perception, Response, Measurement, and Transport

Odor Perception

In the human nasal cavity, odor perception begins within highly specialized receptor cells. According to current olfactory mechanism research, as individual odor molecules are drawn into the nasal cavity, a portion is dissolved in the mucous film that covers these detectors. Once an odor molecule is captured, it attaches to one or more of the individual receptor cells based on a shape match. Depending upon the molecule, it may be captured by one or several of the specifically shaped receptors. Once a receptor has been stimulated, an electrical signal is transmitted to the brain and a most amazing process is underway.

After a signal is received in the brain, a response occurs. The reaction may be to flee because of an association with danger or it may be to linger

Table 12.1   A partial listing of compounds that have been identified in the air above an anaerobic swine manure storage tank

Alcohols	Aldehydes	Ammonia and homologs
Methanol	Acetaldehyde	Ammonia
Ethanol	Ethylaldehyde	Methylamine
1-Propanol	Propionaldehyde	Ethylamine
2-Propanol	Pentanal	Trimethylamine
1-Butanol	Hexanal	Triethylamine
2-Methyl-1-propanol	Heptanal	
3-Methyl-1-butanol	Octanal	**Ketones**
2-Ethoxy-1-propanol	Decanal	Propanone
2-Methyl-2-pentanol	2-Methyl-1-propanal	2-Butanone
2,3-Butanediol	Ethyl acetate	3-Pentanone
		Cyclopentanone
Organic acids	**Cyclic compounds**	2-Octanone
Methanoic acid	Phenol	2,3-Butandione
Ethanoic acid	4-Methylphenol	3-Hydroxy-2-butanone
Propanoic acid	4-Ethylphenol	
Butanoic acid	Toluene	**Sulfur-containing compounds**
2-Methylpropanoic acid	Xylene	Hydrogen sulfide
Pentanoic acid	Indone	Carbonyl sulfide
3-Methyl butanoic acid	Benzaldehyde	Methyl mercaptan
Hexanoic acid	Benzoic acid	Dimethyl sulfide
4-Methylpentanoic acid	Methylphthalene	Dimethyldisulfide
Heptanoic acid	Indole	Dimethyltrisulfide
Octanoic acid	Skatole	Diethyldisulfide
Nonanoic acid	Acetophenone	Propyl mercaptan
Phenylacetic acid	*O*-Aminoacetophenone	Butyl mercaptan
2-Phenylpropanoic acid	Aniline	Dipropyldisulfide
		2-Methylthiophene
		2,4-dimethyl thiophene

Adapted from Miner, 1981.

because of the perceived desirable situation. Humans can detect more than 10,000 different odors based upon the combination of detectors that are activated. Most humans can identify only a small portion of the different odors detected. Our sense of smell is much more precise than is our ability to describe the odor that is perceived. As an indication of the sophistication of the sense of smell, consider the perfume and fragrances industry, which produces a multitude of different fragrances to make our everyday products more appealing or satisfying.

Table 12.2.   Characteristics and barely perceptible concentrations of various substances in air

Substance	Odor characteristic	Concentration causing faint odor, g/1 × 10⁻⁶
Acetaldehyde	Pungent	4
Ammonia	Sharp, pungent	37
n-Butyl mercaptan	Strong, unpleasant	1.4
Carbon disulfide	Aromatic odor, pungent	2.6
Ethyl mercaptan	Odor of decayed cabbage	0.19
Hydrogen sulfide	Odor of rotten eggs	1.1
Methyl mercaptan	Odor of decayed cabbage	1.1
Propionaldehyde	Acrid, irritating	2
Propyl mercaptan	Unpleasant	0.05

Adapted from: Sheehy, Achinger and Simon, Handbook of Air Pollution, PHS Publication 999-AP-44, A17, Updated.

Most natural odors are highly complex mixtures of many different compounds. Humans can discriminate not only between the odor of a peach versus the odor of an apple but also between varieties of peaches, and frequently, trained observers can identify the degree of ripeness of a specific variety. Livestock and poultry odors are similarly complex. More than 100 different compounds have been identified in the headspace over stored manure. Some of these compounds are listed in Table 12.1. Variation in the relative concentrations of the individual compounds may explain the ability of some people to identify the individual farm that has been visited by another person during the day based on the odor residual clinging to their clothing.

It is also important to recognize that there is a complex and often hard-to-explain relationship between the concentration of individual odor compounds and the perceived odor. For example, observers standing downwind of an animal waste odor source are likely to report smelling the odor of ammonia as one component of the odor. When concentrations are measured, it is likely that the actual ammonia concentration is less than the published threshold odor concentration. Two possible explanations exist. Ammonia can be detected in lower concentrations when present as part of a gas mixture or the ammonia smell is actually caused by one of the amine compounds, which is detectible at much lower concentrations than is ammonia. Table 12.2 provides barely detectible concentrations of some of the compounds likely to be found in the vicinity of animal manure. Note that these concentrations vary over several orders of magnitude.

Table 12.3.   Airborne concentration of various gases associated with animal manures under which it is thought that workers may be repeatedly exposed without adverse effects

Substance	Concentration, $g/l \times 10^{-9}$
Acetaldehyde	360
Acetic acid	25
Ammonia	35
n-Butyl acetate	710
Butyl mercaptan	35
Diethylamine	75
Dimethylamine	18
Ethylamine	18
Ethyl mercaptan	25
Isopropylamine	12
Methylamine	12
Methyl mercaptan	20
Triethylamine	100

Adapted from: Threshold Limit Values, American Conference of Governmental Industrial Hygienists, 1967.

Several of the chemical compounds identified in the vapors from animal manure have the potential to cause physiological damage as well as to be part of the odor. In every case, the concentration at which the gas becomes a threat to workers is several times higher than the concentration barely detectible as an odor and to the concentration typically measured in the vicinity of livestock and poultry confinement facilities. Table 12.3 summarizes the concentration of some compounds associated with animal manure at which it is thought workers may be exposed repeatedly without adverse effects.

In addition to the compounds listed in Table 12.1, there are two odorless gases typically associated with livestock and poultry manure that should be considered. Carbon dioxide and methane are both asphyxiants. Methane is also explosive at concentrations between 5 and 15%.

Psychological Response

The psychological response to odors is more complex and less well understood than the physiological response that has been extensively explored during the past 30 years. Evidence suggests that each of us learns to like or dislike certain odors. Children like most smells. It is only as we mature and begin to talk about the odors that we develop a sense of likes and dislikes. Food tastes are very much related to the odors from the food itself. Subtle spicing would be ineffective except for the multitude of odor differences we are able to detect.

Only recently have scientists begun to relate these complex psychological reactions to the ways in which people respond to specific odor sources such as those associated with confinement livestock production. Clearly, people react differently to the smell of any particular odor source. There are experiences of people who react to animal manure lagoons with an emotional intensity that others would find entirely unreasonable. Recent observations suggest these are honest and accurate reactions. Whether these responses are so intense because they have an objection to the odor source based on other factors is unclear. It has been observed, however, that there are fewer objections within a community to the odors that are a traditional part of the community, or that are produced by an agricultural operation of an appreciated and esteemed member of the community compared with an odor generated by an operation owned by an outside agent that may alter the traditional social structure. Thus, a large, high-technology livestock confinement system relocating to an area of traditional-style livestock production can expect to have local residents find the odor more objectionable than an odor of similar intensity from a conventional system of livestock production.

Recent research indicates there are some specific physiological changes that can be measured in the endocrine systems of persons living downwind of a large-scale livestock operation. These changes result in more frequent feelings of helplessness and depression. There is a sense among many of these residents that they cannot plan an event for guests at their home because of the uncontrollable possibility that the wind direction might be such that their home would be filled with a livestock manure odor.

Odor Intensity Measurements

Odor intensity is usually one of the primary factors in determining whether an odor problem exists. Several methods of assessing odor intensity have been proposed and used. The most obvious method is the direct scaling technique. This approach involves asking panelists or other unbiased observers to evaluate an odor intensity on a numerical scale. Some researchers have proposed scales of 1 to 6; others have used 1 to 10 with the higher numbers representing the more intense odors. This technique was used as long as 60 years ago. A modification of this technique, which improves its usefulness, provides panelists with a standard of defined intensity. This tends to reduce variability among panelists.

Referencing is an alternate technique. A panelist is asked to compare the intensity of an unknown odorant with a series of different concentrations of a reference odorant. This panelist indicates whether the odorant being evaluated is more, less, or of the same intensity as the standard. The most common standard is 1-butanol because of its availability in a highly purified form, its low toxicity, high stability, and reasonably agreeable odor. Olfactometers based on 1-butanol have been constructed by researchers for the past 25 years. More recently, portable units have been constructed and techniques perfected for bringing odorous air samples into the laboratory. Exposing cotton fabric swatches to odor sources and then comparing the odor of these swatches also has been proposed as an alternative to either transporting the odorous air to panelists or to transporting panelists to the odor source. Odor researches seem to agree the odor absorbed by a clean, dry cotton swatch is of the same general quality as the odor at the site where it was exposed.

Dilution, either liquid or vapor, has been used as an approach to evaluate odor intensity. Although several techniques have been used, all are similar in concept. Panelists are presented samples of diluted odors to determine the dilution of the odor they can barely distinguish from an odorless sample. Both liquid and gaseous comparisons have been conducted. The equipment for these comparisons ranges from the relatively simple to the highly complex. The Scentometer, a commercially manufactured device, can be taken to the field and diluted samples of the ambient air compared with odor-free air. Scentometer evaluations of odor intensity have been widely used because of the low cost and convenience of the tool. More precise dilution olfactometers are in widespread use for research purposes. The devices are all similar in that they deliver diluted samples of odorous air to panelists for comparison to an established standard.

An indirect approach to measuring odor intensity has been to measure the concentration of a constituent that is present in the odorous air and is easily measured. This approach has particular appeal if the constituent being measured is a major contributor to the odor being represented. Ammonia and hydrogen sulfide have been the two constituents most commonly measured in manure odors. Volatile fatty acid concentrations also have been used as a surrogate. Although most researchers seem to agree that none of these fatty acids is the major constituent of odors from a livestock or poultry facility, their removal or reduction in concentration has frequently proven to be a useful alternative measure of odor reduction.

More recently, electronic sensors have been proposed that might have the potential for odor intensity measurement. If such a device was available, it would contribute a degree of objectivity to the odor measurement process that is difficult to achieve with current measuring techniques. Most electronic detectors currently available respond to a particular constituent or a small number of constituents so they have not proven appropriate for an odor as complex as that from an anaerobic manure storage. A broad-range electronic detector has been manufactured and was proposed as an odor-measuring device, however, initial testing at cattle feedlots and dairies has not been encouraging.

Regulatory Efforts to Control Odors

Livestock and poultry production odors are not regulated by federal statutes or by the state air pollution control agencies. Many states, however, have expressed concern over the number of complaints they have received from citizens related to odors from confinement livestock production. Because of the difficulty in defining the frequency, intensity, duration, and offensiveness (FIDO) factors, regulations have been difficult to formulate and would be equally onerous to enforce. As an alternative, several states and local units of government have established regulatory guidelines concerning the location, design, construction, and operation of confinement facilities that are hoped to indirectly reduce the number of odor complaints. One of the approaches has been to specify the minimum required separation distance between confinement buildings or feeding pens and nearby residences or public gathering areas. A second has been to regulate the use or location of anaerobic lagoons. A third approach has been to specify a minimum area for manure disposal to be located in proximity to confinement systems. A fourth approach has been to insist on adoption of best-available odor control technology (locally defined) as a condition of issuing a construction permit. Most areas do not currently have a regulatory process that specifies the extent to which odors are to be controlled, nor do they have regulations that adequately protect the livestock or poultry producer from odor complaints or more serious legal action.

Several states have adopted "Right to Farm" legislation designed to protect agricultural pursuits from the encroachment of housing or commercial developments into areas of established agriculture. The goal was to avoid lawsuits that were resulting from having persons unfamiliar with agricultural odors attempting to halt or change agricultural practices after moving into the area. The provisions of the Right to Farm laws are different and the protected practices vary from state to state. Most protect the opportunity to spread manure on cropland. Others provide an establishment time. Any agricultural practice that has been continuing for the specified time or longer is protected from the arrival of new residents or the development of new concerns by existing neighbors.

The greatest threat to confinement livestock producers relative to odors is the private or public nuisance suits that have been filed at various locations around the United States. Without specific regulations regarding acceptable odor levels, offended residents have recognized private or public nuisance litigation as their final source for achieving a remedy. Nuisance litigation is based on the concept that no one has the right to unreasonably interfere with another's right to enjoy his or her property. When present in unreasonable concentrations or for unreasonable periods of time, odors have been considered sufficiently important to be judged a "legal" nuisance on several occasions. Each case must be considered separately and the judge and jury generally evaluate each location and situation separately in deciding whether the plaintiff has suffered an unreasonable nuisance situation, e.g., an unreasonable odor.

The burden of proof of the nuisance falls to the plaintiff. Where a nuisance has been judged to exist, damages may be assessed against the odor source. Damages may be actual for such matters as medical costs, alternate housing, decreased property values, or more frequent cleaning of furniture or draperies. Punitive damages also may be assessed where it is decided that additional payment is necessary because the damages were caused by negligence or irresponsible actions. Nuisance lawsuits have been heard in almost all of the major livestock-producing states and have proven to be an expensive alternative to odor control for producers who have become directly involved. Court decisions have been made both in favor of the plaintiff and in favor of the livestock enterprise when the judge and jury were not convinced that a nuisance condition had been created.

Climatology and Odor Transport

Wind direction is important to odor transport. Traditionally, it has been suggested that livestock and other odor-generating activities should be located downwind of sensitive sites. Although sound advice, wind tends to blow from several directions at any particular location. Being downwind according to the prevailing wind may be helpful but does not afford complete protection. Wind direction data are typically presented as wind roses. There are several representations available; Figure 12.1 is an example. Note that this wind rose presents not only a record of the wind direction data but also includes information on wind speed. Figure 12.2 is an example of a simpler wind rose showing only a distribution of wind directions.

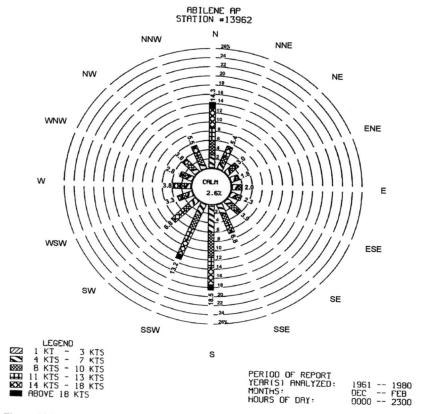

Figure 12.1. Example wind rose. This presentation is for a particular weather station and for a particular season. It represents a summary of several years of data. The fraction of time the wind is blowing from each of the 16 directions can be read. The legend allows insights as to the distribution of wind speeds from any of the directions. The fraction of calm is shown in the center of the rose.

Odors released into the environment are transported by wind and are subsequently diluted by atmospheric turbulence. Turbulence increases with greater wind velocities and with the presence of surface roughness such as hedges, trees, and buildings. Incoming solar radiation warms the earth's surface. Some of this heat is transferred to the overlying air, which expands as it warms, creating an upward buoyancy. This unstable condition, lighter air beneath heavier air, contributes to the normal process whereby odors are dissipated in the environment. The Gaussian plume dispersion equation has been adopted as the most common starting point in attempts to predict the extent and concentration of

1961 TO 1988 AUG

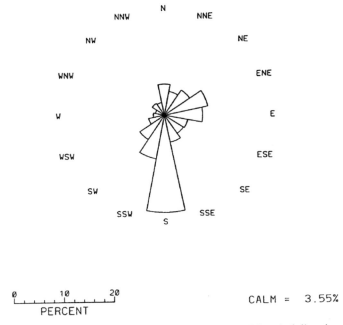

Figure 12.2. Simple wind rose that shows the distribution of the wind direction among the 16 components. The fraction of calm is shown outside the rose.

downwind odor concentrations. This highly theoretical model has the advantage of being amenable to mathematical solution, but in exchange it ignores many of the local topographical features that are important in understanding odor transport.

Both modeling and practical experience have indicated that the most intense odor concentrations occur under highly stable atmospheric conditions known as temperature inversions. Temperature inversions occur when the air temperature increases with height. Such a ground inversion can be noted on a clear, wind-free night when there is rapid cooling of the earth's surface and by conduction, a cooling of the air in contact with the surface. As wind speed increases, typically in the morning, mechanical turbulence begins to dominate the odor transport process. Climatic conditions contributing to greatest downwind odor concentrations are noted in Table 12.4.

Table 12.4. Climatic conditions contributing to increased downwind odor perception

Climatic condition contributing to increased downwind odor detection	Climatic conditions tending to dilute downwind odor concentrations
Low wind velocities	Higher wind velocities, >10 mph, contribute to more rapid dilution and dispersion of odorants
Warmer temperatures cause greater odorant volatility	Cool temperatures slow odorant escape
Presence of a temperature inversion trapping odorous air beneath a dense overlying layer	Normal temperature decrease with elevation, allowing warm air to rise
High relative humidity causing odorants to cling to water particles and increasing the ability of the nasal passage to detect odors	Low relative humidity
Recent rainfall on manure-covered surface such as earthen feedlots	Recent dry weather causing manure surfaces to be dry
Wind direction transporting odors directly toward sensitive neighbors	Wind blowing odors away from most sensitive neighbors

The combination of air velocity, temperature change, and terrain can cause unexpected odor transport. For example, during a clear night with low wind velocity, cold air after being in contact with an odor source can flow down a hill and fill a valley with odors that dissipate much more slowly than would otherwise be expected. Although it is always wise to consider prevailing wind directions in the selection of a livestock facility site, it also must be remembered that the wind in most parts of the world blows from all possible directions for a part of the year. As a result, there is no direction that can be considered immune from odor transport.

Models based on the Gaussian plume calculations to predict the movement and dispersion of animal waste-related odors have been used in the Netherlands and are sometimes required in Australia. They have not been applied to any extent in the siting of livestock and poultry facilities in the United States because of the problems inherent in the process. Standard Gaussian plume models are not designed to simulate atmospheric pollutant transport over rough terrain or around irregularly shaped objects, nor do they account for the effects of vegetation. Most severe odor invasion episodes occur when wind velocities are either low or the wind is stagnant. These are the conditions under which modeling is least likely to be an accurate representation of reality. Figure 12.3 is a representation of odor plumes that would be predicted by the Gaussian model under a number of climatic conditions. These representations assume a smooth horizontal surface, a point source of odor, and no interactions other than dilution and turbulent transport.

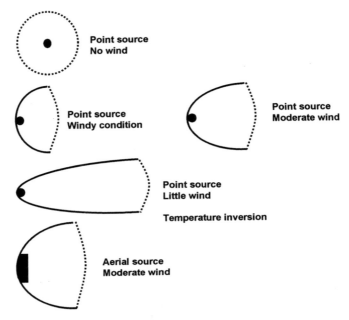

Figure 12.3.   Downwind odor distributions as predicted by the Gaussian model under a number of climatic conditions. Note that the greatest transport distance is predicted to occur under conditions of low wind speed and when a temperature inversion exists. The temperature inversion tends to restrict vertical mixing, hence causing a larger downwind odorant concentration.

Odor Sources Associated with Livestock and Poultry Confinement Facilities

Odorous gases can be produced at several sites around a livestock enterprise. The most common odor sources, however, are the floor and other surfaces of buildings and pens, the bodies of the animals, manure storage and collection facilities, feed storage facilities, dead animal disposal and storage areas, and manure exposed to the air during land application. Each of these odor sources has received research attention, and technologies exist that can be used in response to the odor issue. One of the currently unresolved issues is the need for odor control and the associated issue of the willingness and capability of livestock and poultry producers to invest in odor control. Certain odors have been historically associated with livestock production and regarded as "normal." Producers and their neighboring landowners accepted these odors as part of their environment. As livestock enterprises have grown in size and become more specialized and as rural residents have become less tolerant

Table 12.5.   Items worthy of consideration in
evaluating a potential site for the construction of
a livestock confinement facility

Check list for site selection
Topography
Air movement
Facility size
Soils and geology
Soil–plant filter quality
Management level
Wells and streams
Visibility of facilities
Property line location
Neighbors' location and attitude
Potential for additional development
State and county roads
Public use or access areas
Waste management facilities

Adapted from Charles Fulhage, Coy McNabb, and
John Rae. Confined Feeding Facilities: Site Selec-
tion and Management, University of Missouri-
Columbia Extension Division, Undated.

of invading odors from nearby enterprises, the need to address odor
management issues has become acute.

The single most important aspect of odor control is that of site selection. A properly selected site is one with adequate land for manure use
and one that provides adequate buffer distance from neighboring residences and alternate land uses that are likely to be sensitive to livestock
or poultry odors. Hospitals, schools, parks, golf courses, churches, housing developments, communities, and commercial establishments such as
restaurants and motels are highly sensitive to odors. Additional factors
worthy of consideration in selecting a site for a livestock or poultry facility are listed in Table 12.5. Rural residences located near the facility are
also likely to find that the odors interfere with the normal use of their
property. Several groups have suggested separation distances that in
their judgement were appropriate. Examples of suggested separation
distances published by the University of Missouri-Columbia Extension
Division are listed in Table 12.6.

Odors from Unroofed Feeding Areas and Their Control

Odors from unroofed beef cattle feedlots and from outdoor dairy cattle
feeding pens are frequently matters of concern. In addition but to a

Table 12.6.   Guide for estimating separation distances (values in the table are for estimation only; distances greater than those shown do not guarantee success, and distances less than those shown may not guarantee failure)

Number of finishing hogs or beef cattle	Fair to good separation distance, ft	Suggested distance from nonowned residences, ft	Suggested distance from residential development or public recreational areas, ft
500	500	1,000	2,000
1,000	750	1,500	3,000
1,500	900	1,800	3,600
2,000	1,000	2,000	4,000
2,500	1,200	2,400	4,800
3,000	1,500	3,000	6,000

Adapted from Charles Fulhage, Coy McNabb and John Rae. Confined Feeding Facilities: Site Selection and Management, University of Missouri–Columbia Extension Division, Undated.

lesser extent are concerns over odors from outdoor pig feeding operations and poultry raised on open lots. The reason the pig and poultry facilities have been of lesser concern is probably related to the generally smaller size of these operations and their being less numerous than beef cattle feedlots or open-pen feeding of dairy cows. Because cattle feedlots are the most numerous and have been the cause of more widespread concern, they are addressed as the representative odor concern situation. Transferring cattle feedlot odor control concepts to other species raised in unroofed confinement is a direct transfer.

Cattle feedlot odor control is sufficiently related to both dust and fly control that they, too, must be considered. Most odor complaints related to feedlots involve all three. Many of the design and management decisions to reduce odors also would be effective in controlling fly populations. Dust control is also necessary if odor control is to be considered successful because cattle feedlot dust carries odor with it. The airborne particles that are identified as feedlot dust include dry manure. There is also the challenge of not worsening the odor problem by the application of poorly planned or executed dust control measures.

Source of feedlot odors. The most frequent source of cattle feedlot odors is that arising from the anaerobic decomposition of manure on the feedlot surface. When feedlot surfaces are wet (overall moisture content greater than 70%), there are major wet spots within the pen; animals are covered with mud and manure; cattle have access to standing water in low areas or pools; or runoff-transported manure is allowed to stand in uncontrolled pools around the facility, manure decomposition by-products are released that are detected by surrounding residents or persons traveling through the area as an objectionable odor. Dead animals allowed to remain on the premises exposed to air also produce both an undesirable odor and an unacceptable sanitation issue.

The runoff control facility of a feedlot is also a potential odor source. The solids settling facility is subject to odor production if the liquids are not drained away in a timely fashion. The runoff retention basin if allowed to remain filled for an extended period also may contribute to odor; however, the surface of the retention basin is much smaller than the feedlot surface, hence, its overall contribution to odor is generally small. The land application of retention basin contents is an additional odor source, but one that can be minimized by appropriate planning and scheduling of the pumping. When pens are cleaned and the manure pack disturbed, odors are likely to increase.

Perhaps the most challenging odor issue is that created by the feedstuffs that are stored on site. In many areas of the United States, cattle feedlots have developed in the area of vegetable processing operations that have one or more waste products suitable as cattle feeds. The potato processing of the northwestern United States is a prime example. Potato processing waste is generally formulated as a slurry that is available in relation to the processing schedule. Feeding this material to cattle solves a major disposal issue for the processor and provides a low-cost energy source for the cattle. The challenge lies in that the shallow pits typically used for the storage of potato waste slurry are significant odor sources. Additional odor is released as the potato slurry is removed from the pits, mixed with the other ration ingredients, and distributed into the feedbunks. Other food processing wastes also are fed to cattle and raise similar odor control issues but are less prevalent than potato waste feeding.

Feedlot design to minimize odor release. The best opportunity to minimize feedlot odors arises during the site selection, design, and construction phase. It is important to select a site that provides adequate land for the feedlot and anticipated growth as well as adequate land for runoff and manure disposal. In addition, separation from surrounding incompatible land uses is important. The separation distances for feedlots are similar in magnitude to those for swine operations but again, no specific separations ensure there will not be conflicts. Feedlots with animal capacities in excess of 20,000 generally try to have a buffer of 1 mile from sensitive alternate land uses. Geographical regions with low rainfall and rapid drying conditions generally accommodate feedlots with fewer odor complaints than those where the feedlot surface remains wet for a longer period.

A southern- or western-exposed slope to a feedlot site also aids in rapid drying of the surface. Concrete-surfaced pens afford considerable control over the development of low, poorly drained areas within a feeding pen. Where complete concrete surfacing is considered too expensive, a frequently selected alternative is to surface the 10 feet adjacent to the feedbunk because this is an area with more animal traffic than the remainder of the lot. Similarly, the area around the watering device is one that tends to stay wet much of the time and can hold moisture if not surfaced. If a major part of the feeding pens is unsurfaced, it is important that it be sloped sufficiently to drain well. Minimum slopes of 5% are helpful. Where lots are built with slopes less than 5%, one alternative is to build mounds within the lots to provide the animals a dry place to stand and prevent the development of water-holding depressions elsewhere in the lot.

Fence lines and bunks are additional areas that can contribute to odors and to fly breeding. Feedbunks should be designed so they do not create areas beneath the bunk where wasted feed accumulates or where manure can be pushed to create a slow-drying area that does not get sufficient animal traffic to promote drying. Fences should be constructed so there is minimal accumulation of manure along the fence line. The fence also should be designed so a scraper can be used to move any manure that accumulates back into the pen area. It is also important that when feedlots are cleaned, typically once or twice annually, that the seal that develops at the soil–manure interface not be disturbed. This is typically a very well compacted area that effectively limits downward movement of water and dissolved nitrogenous compounds.

The runoff control system is a second area in which design can help avoid odor generation. There should be a surface drainage system that channels runoff out of the lots and allows it to flow toward the runoff retention basin. Typically, there is a solids settling basin included in the system between the lot area and the retention basin. The solids settling facility is typically a section of the runoff channel system that is widened and flattened so the flow velocity is reduced sufficiently for solids deposition. Ideally, this area should drain within 2 days of a runoff event and within a week to 10 days, it should be sufficiently dry such that the solids can be removed with a front-end loader or other solids handling device.

Feedlot management to minimize odors. Assume a feedlot was designed with odor control included as a priority consideration. The management challenge is to maintain the facility such that all of the design provisions continue to function. Thus, the feeding pens should be groomed as needed to prevent low spots where moisture could accumulate. Any manure or wasted feed that accumulates along the fence line or feedbunk should be removed. Runoff diversion channels should be maintained to allow runoff to move promptly from the lot. In addition, animal densities could be manipulated to control lot surface moisture content. During hot weather, it is appropriate to increase animal loading rates to take advantage of the additional urine as a dust control feature. During cold or rainy weather, it is appropriate to reduce animal densities to aid drying of the lots that otherwise may become wet and sloppy. Wet, sloppy lots not only contribute to elevated odor conditions but also interfere with maximum rates of gain.

When removing accumulated runoff from the runoff retention basin, it is important to minimize the impact on neighboring landowners. Assuming an irrigation system is being used, it is reasonable to consider wind direction and to spread the liquid early in the day to achieve the most rapid drying rate. If a center pivot irrigation system is being used, low-pressure nozzles that discharge as close to the ground as possible reduce odor transport. Whatever type of irrigation system is used, it is important that the application rate be less than the infiltration rate of the soil to avoid runoff. Runoff water should be applied in amounts consistent with the nutrient management scheme developed to avoid over-application and potential groundwater contamination.

Control of odors from feed storage facilities. Certain cattle feedstuffs increase odors in the vicinity of the feed storage, mixing, and feedbunk areas. Vegetable processing wastes such as those from potato or fresh vegetable processing plants are suitable for incorporation into cattle feeds, but if stored in pits or trench silos are an additional odor source. Controlling the escape of these odors may be difficult. The surface area can be minimized to the maximum extent possible. Some wastes can be handled in a sufficiently dry form so as to promote a dry surface that reduces odors. Covers may be practical in some situations; in others, it may be necessary to construct off-site storages if the feedlot site lacks sufficient separation to allow feed storage without objectionable odors.

Dust control around feedlots. Dust control from cattle feeding areas has arisen as a significant challenge at many feedlots, particularly in the more arid regions of the United States. Dust is particularly likely in the early evening when the temperature cools from the midday high and the animals become more active. Dust control is generally achieved by increasing the moisture content of the lot surface. The first and perhaps most popular response to dusty lots is to increase animal densities to take advantage of the additional urine supply on an aerial basis. This approach is likely to be effective in the cooler, less arid areas but is generally insufficient to control dust in hot, arid regions during the peak dust season. If a dust problem cannot be sufficiently managed by adjusting animal densities, the next alternative is to apply water with either a water truck equipped with a large spray nozzle or to install a sprinkling system. The challenge is to apply water on a frequent basis and sufficiently uniformly that dust is controlled throughout the lot but so that no areas of the lot receive enough water to become sufficiently wet to promote odors due to anaerobic fermentation. Achieving this end generally means installing specially designed sprinkler systems that cover the entire lot and having the lots sufficiently well groomed that wet areas do not develop. Application rates of between 0.25 and 0.5 inch per day are generally sufficient.

Roadways and cattle working alleys are additional dust-generating possibilities. It may be possible to control dust in these areas with mobile water-hauling equipment. Controlling dust contributes to improved working conditions for the feedlot staff and helps avoid complaints from neighboring residents.

Pen cleaning may contribute to increased odor and dust problems. Fortunately, this cleaning period is generally brief and it may be possible to endure the situation by carefully selecting the times of pen cleaning and application of water.

Fly control around feedlots. Flies are a potential nuisance around feedlots both to neighbors and the feedlot staff. In addition, if fly populations become sufficiently large, they interfere with animal performance and their control becomes an economic as well as an environmental issue. Currently, fly control is best achieved by preventing the development of breeding and propagation areas. Active feedlot surfaces in which animals are confined at typical loading rates do not allow fly breeding due to hoof action. Flies breed in warm, moist manure that is allowed to stand undisturbed. These areas tend to develop where there are low spots in lightly loaded feeding pens, along fence lines, or adjacent to feedbunks where feed is spilled and not collected or trampled. Thus, fly control is best achieved during design when these conditions can be avoided in the facility to be constructed.

The most severe situations are those in which an older, poorly designed and constructed feedlot has been allowed to expand without improving the old lots or where a small number of animals are left in a pen after the bulk of the animals is sold. Thus, housekeeping and maintenance are general keys to preventing fly population explosions. Solids settling basins, if they do not drain well, may harbor fly breeding sites. In this case, it is appropriate to modify the settling area to achieve more rapid dewatering and to remove the accumulated solids before the fly larvae hatch.

Insecticide spray programs and fly traps are available for use around feedlots and are generally used in conjunction with manure handling procedures to limit breeding areas. Most operators have found that it is not feasible to control flies by spraying or trapping without first minimizing the amount of wet manure in which breeding occurs.

Fly nuisance conditions also can be reduced by minimizing the fly harboring areas such as tall weed and brush growth in the vicinity of feedlots. Several complaints have been lodged concerning flies that were eventually resolved by eliminating areas in which flies were resting between the feedlot and the neighboring residence from which the complaints arose.

Odors from Enclosed Confinement Buildings with Liquid Manure Handling Systems

Odors from the floor, pen partitions, and bodies of confined animals have been a serious issue that has been reduced in occurrence as designers become better informed about how to design buildings that promote clean animals and pens. The use of slotted floors and flushing gutters have simplified separation of animals from their manure. Early confinement buildings reflected a limited understanding of animal physiology and ventilation needs. As a result, dirty, manure-covered animals were not uncommon. Pigs were forced to roll in manure as a source of cooling. Dirty animals were a particular odor problem because of the additional area of manure-covered surface for odor generation and because the animal body was a warm surface promoting decomposition and vaporization. Contemporary building design ensures that animals are maintained at a comfortable temperature and that the design of the pen provides an opportunity for the animal to remain clean, dry, and manure-free. Misting equipment has been added to swine buildings in warm climates to enhance evaporative cooling of animals. By minimizing the amount of manure on pen floors and along partitions, the production of odorous gas is reduced. Clean animals and pens can be achieved in a variety of ways. The use of properly ventilated pens, appropriate animal stocking densities, and either partially or fully slotted floors achieve the desired result. Open flushed gutter systems also have been effectively used.

The manure collection, storage, treatment, and transport system has been the focus of odor control research because of the earlier observation that this is the odor source of greatest importance to down-wind neighbors. Storage of manure beneath the floor and outside the building in storage structures is widely used. The underfloor storage tanks are generally designed for a minimal water addition to extend the storage capability. Most underfloor storages are allowed to stand quiescent. One exception has been the use of underfloor oxidation ditches that were popular in the 1970s. With adequate mechanical aeration, the storages can be maintained at markedly reduced odor levels. In addition, much of the nitrogen lost was by alternate nitrification–denitrification. Although the aeration of underfloor manure storage tanks improved the quality of the building environment, energy costs were sufficient to make this alternative less attractive to livestock producers by the 1980s. The more typical approach to underfloor storages has been to install a portion of the ventilation capacity so that it draws air from above the liquid manure level and below the slatted floor to reduce the upward movement of odorous gases.

Another popular approach to handling the manure from confinement buildings is to remove it from the building for storage and treatment. Alternatives for manure removal include a variety of scraping devices that operate in shallow gutters, flushed gutters that are located beneath slotted floors or that are open to the animals, and short-term storage pits that are partially filled with water and then dumped on a regular basis (so-called pit–drain systems). The scraping devices involve considerable mechanical maintenance and have been blamed for some of the elevated building odor levels because of the residual manure that is left after the scrapper passes. Both open and underfloor flushed gutters offer the possibility of removing manure on a daily or more frequent basis, the use of fresh water or recycled effluent as the flushing medium, and a system that is easily automated. The pit–drain systems also are easily adapted to the use of recycled effluent. Odor can be reduced if the pits are emptied before significant waste decomposition occurs. Recommendations are to empty these pit–drain systems at least weekly.

Exterior manure storages may be either above or below ground. Aboveground storage tanks are generally constructed of concrete or steel coated with a protective corrosion-resistant coating. Manure is pumped into these tanks from a receiving pit with some type of open impeller centrifugal pump. The belowground manure storages may be in a watertight tank made of concrete or some other impervious material. The tank may be covered or partially covered to reduce the rate of air exchange between the manure and the ambient air. Odor regulations are being developed that require covers on manure storages and storage tanks for recycled lagoon liquid to be used for flushing. There are also earthen manure storage tanks that are constructed in compacted high clay-content soil to minimize infiltration. They are generally built with protective berms to prevent the entry of runoff water. Manure storage tanks are designed to operate with minimal water addition. The tanks are generally not agitated until immediately prior to emptying as a way to reduce odor release. Agitating a previously quiescent manure storage tank releases a large quantity of odorous and potentially lethal gases. Odor release from manure storage tanks can be reduced by keeping the tank surface covered. Most odor complaints occur when the tanks are being agitated or emptied. If manure is being loaded into a trailer or truck-mounted tank, agitation and splashing can be minimized by proper equipment selection and placement.

Safety Note

As noted in Chapter 5, the gases released during the agitation of manure storage tanks are sufficiently concentrated and in sufficient quantities to be lethal to people who might be standing adjacent to the tank. Operators should protect themselves, their families, and their employees by having them well away from the manure storage during agitation. The gases that have accumulated during storage can lead to both suffocation and actual toxicity.

Controlling odors from anaerobic lagoons. A second popular manure management system includes the incorporation of an anaerobic lagoon. Lagoons provide a low-cost means of manure storage and achieve a high level of nitrogen- and oxygen-demand reduction. Lagoons are also a frequent source of odors. Furthermore, lagoons result in a large portion of the nitrogen originally present in the manure being released into the overlying air as ammonia. Proper lagoon design (appropriate organic loading rates) and management are intended to achieve a relatively low-intensity odor release. This approach is effective much of the time, however, there is a sufficiently strong level of concern that lagoons have been eliminated as an alternative in some areas of the United States. Odors from lagoons are particularly severe during late spring and early summer when the water temperature increases to the point that biological activity becomes more intense, converting the manure that accumulated during the winter into intermediates that are more odorous than when the operation stabilizes. Because lagoons have been constructed to accept the wastes from large swine facilities, the intensity of public objections has been sufficient to prompt consideration of alternative treatment processes.

One method that is widely credited with reducing the odor intensity of anaerobic lagoons is to reduce the loading rate, or, to increase the treatment volume relative to organic loading. Lagoon loading rates are frequently expressed as lagoon volume per unit weight of animal served. Recommended volumes range from 1.5 to 3.0 cubic feet per pound of animal served for swine in the United States. Lagoons in the warmer regions can be sized at the higher loading rates because the warmer water temperatures promote more rapid biological breakdown.

Chemical and biological additives, masking agents, and other proprietary products are available for use in anaerobic lagoons and other manure storage structures. Typically, they are sold to improve lagoon performance, liquefy the accumulated solids, and reduce odors. There is little supporting data to document the success of these materials. There is generally no shortage of well-acclimated bacteria in the lagoon nor is there a problem with pH in a properly designed and operating lagoon.

In addition to an adequately sized primary anaerobic lagoon, a lightly loaded, second-stage lagoon is often included that provides further treatment and storage of effluent. This lagoon usually results in effluent of lower odor potential for land application or recycling as flushwater.

Two types of lagoon covers have been proposed for odor control. Impervious covers, rubber or plastic material with fiber reinforcement, have been installed on several industrial waste, slaughterhouse, packing plant, and food processing anaerobic lagoons. They also have been installed on a few anaerobic lagoons serving livestock confinement facilities. They have proven effective but have not been more widely adopted, in part because of the cost. The captured gases can be burned or can be deodorized by venting into a soil bed. Researchers have noted that when a floating scum develops on dairy manure anaerobic lagoons, odor release is reduced. A synthetic, floating permeable cover or floating balls simulating a natural floating scum have been developed but are not commercially available. Some operators have spread straw, peat, or similar material on lagoon or manure storage basin surfaces to reduce odor escape. These permeable covers operate in two ways: they provide an aerobic absorption and treatment layer for permeating gases and they protect the liquid surface from wind.

Alternatives to anaerobic lagoons. Aeration is an alternative that is available to designers of lagoons and other storage units. By the addition of mechanical aerators, the anaerobic process can be converted into an

Figure 12.4.   A typical aerated lagoon shown with a floating surface aerator to maintain a measurable dissolved oxygen concentration in the surface layer.

aerobic one with the associated odor control benefits. A typical mechanically aerated lagoon is shown in Figure 12.4. Most often, when an aeration process is selected, a portion of the organic loading is removed by some other process, either sedimentation or screening, to separate the solids or anaerobic digestion in an enclosed environment. Thus, one possible manure handling system with a lower odor potential would be the use of a solid–liquid separator followed by an aerated storage lagoon. Another option would be a covered anaerobic lagoon followed by an aerated lagoon.

Anaerobic digestion with biogas recovery has been practiced in much of the world as a way to reduce odor release and to claim a portion of the energy as biogas fuel or as locally generated electrical energy. These systems are widely used in Asia but have had limited appeal in the United States because of the higher operating costs and the greater initial investment. However, increased attention is being directed to covered lagoons to control odors and to reduce organic constituents and oxygen demand. The technology is available for the construction of manure handling equipment of this type.

Odors from Enclosed Confinement Buildings with Solid Manure Handling Systems

Conceptually, solid manure handling systems have the potential to produce less odor than liquid systems. If by adding bedding, reducing water use, or evaporating sufficient water the moisture content could be maintained at 60% or less throughout the storage, aerobic conditions could be maintained. In reality, it has proven difficult to maintain the manure in a sufficiently dry condition to achieve this goal.

Perhaps the operations that have come closest to maintaining an acceptable odor level are the horse stables around the country in which recreational horses are maintained for use by their owners. Horses are typically expensive both to purchase and to maintain, hence they are provided an adequate supply of straw, wood shavings, or sawdust bedding that is replaced on a frequent basis so the horses remain clean. The used bedding is therefore a relatively dilute manure in a bedding mixture that easily composts upon removal from the stalls or it can be used immediately as a mulching material around ornamental plantings. Mushroom producers also seek used horse stable bedding and manure to compost as a medium for edible mushroom production. When composted in the magnitude typical of mushroom producers, there are some odor complaints, but the horse manure bedding operations are relatively free of odor complaints. Small operations for prized show or breeding animals also may maintain bedded housing areas.

Poultry raised on litter. Somewhat related to the bedded raising of large animals is the confinement poultry operations that maintain birds on a litter-covered floor. Typically, litter consisting of sawdust, wood shavings, peanut hulls, wheat, straw, or other absorbent material is spread 4 to 8 inches in depth on the floor of a confinement building before the birds are introduced. The dry bedding is available to absorb moisture from the droppings and to allow air to permeate the bedding and the droppings; thus, it maintains an environment healthful to the birds as well as one pleasant to the caretakers. In addition, it is not offensive to neighboring landowners. Between flocks of birds, it is typical to enter the building with some type of tiller and stir the litter to promote composting. Additional litter may be added if necessary. The used litter is replaced with fresh material, typically once or twice annually. The used litter may be land-applied immediately, composted, or stored in a stack area for use at a more appropriate time.

Although this process is relatively straightforward and would seem to be of a low odor-producing nature, operation has not been without problems. Maintaining the litter on the floor at an appropriate moisture content is difficult. If the litter is too dry, the building tends to become dusty, bird performance deteriorates, and neighbors are bothered by the dust that is carried downwind into their living areas. If the litter becomes too wet, whether due to a leaking water system, invading soil moisture, or inadequate drying conditions within the building, the litter begins to evolve ammonia and other reduced gases, thereby increasing the odor level. In addition, the elevated ammonia concentrations adversely impact the birds that are inhaling air in immediate contact with the litter. In many areas, it is difficult for a manager to cope with inadequate drying conditions because the only means of temperature control may be to reduce the ventilation rate and retain more of the heat from the birds. This response frequently results in house conditions detrimental to birds and workers.

Under ideal conditions, it should be possible to maintain the litter at the appropriate moisture content to avoid elevated odors. Practical experience has demonstrated that leaking watering devices, wet soil conditions, or sustained periods of high humidity and low temperatures are sufficiently frequent to suggest that provision to account for these problems is appropriate. The ability to add supplemental heat to improve drying conditions is one useful alternative. Being certain that moisture is not drawn into the litter layer from below is another helpful step. Placing additional litter on the floor also provides some temporary relief. The removal of spent litter and its storage, management, and disposal are other sensitive processes. Litter hauled in open trucks and trailers has been notorious for blowing onto neighboring properties. Used litter stored outside the building but unprotected from rainfall has been an additional problem. Covered storage areas and covered transport vehicles can resolve either of these issues.

The challenge of odor control for these particular enterprises may be related to the highly competitive nature of the business and the small margins of profit available for taking the additional steps necessary to achieve low odor levels. If this is the case, additional improvement will come as the public and industry recognize that the cost of manure management is a legitimate part of the production cost.

Deep pit poultry confinement. A popular alternative for the confinement of layers in egg farms is to maintain the birds in cages for from two to five hens per cage. The cages may be on a single level or they may be in a stair-step arrangement (Figure 12.5). Whichever cage arrangement is used, the bird-housing area with its associated feed and water distribution systems is located 10 or more feet above a droppings storage area.

Figure 12.5.   A typical arrangement of poultry cages over a deep pit for dry droppings storage.

This system, popularly called the deep pit system, is designed on the basis that at least a portion and frequently all of the ventilation air is exhausted from the manure storage area. The design of a building of this style is based on there being enough ventilation air passed over the accumulated droppings to maintain them in a sufficiently dry condition to inhibit anaerobic decomposition. In areas of extended or severe cold weather, the successful operation of a deep pit building may require more supplemental heat being added than would be necessary to maintain the appropriate temperature for the birds.

Like the litter-based systems, deep pit poultry buildings have the potential for low levels of odor emission but these levels can only be achieved if the droppings are indeed kept dry. From a design perspective, therefore, no water can enter the storage area either from wet soil conditions or from surface water inadvertently entering the building. Also, the watering system must not leak. One of the ongoing challenges in the poultry industry continues to be the design and manufacture of a nonleaking watering device that is sufficiently reliable for installation in a deep pit manure storage system. A low-odor deep pit system also requires that there is sufficient air handling and heating capacity to dry manure even when it is cold and rainy. Any astute egg producer knows it is much cheaper to turn down the ventilation system when it is cold and turn it on to full capacity when the weather is hot. With a deep pit system, it is important that the pit be ventilated even when additional fuel must be used to maintain the birds at full productivity.

In addition to proper design, a deep pit also requires an astute degree of management both to check for water leaks that could cause wet spots in the droppings and to ensure that discovered leaks are repaired immediately. It also requires a high degree of commitment to keep ventilation at the needed rate even though it is so cold that supplemental heating is required. Most deep pit buildings are cleaned on an annual basis, but some have been designed to operate 2 or more years between cleaning. They are typically cleaned by opening a large access door directly into the storage area and by using a front-end loader or other compact solids handling vehicle to move the accumulated droppings from the storage to a waiting truck. From there, they are transported to a land site for soil incorporation or to a site for further treatment prior to reuse. Well-dried poultry droppings are a rich source of plant nutrients and should be covered to prevent being blown from the truck.

In the deep pit system, fly control depends on the success of odor control. If accumulated droppings become wet, they are not only a major odor source but also a source of extensive fly breeding. An associated management commitment is that dead birds do not belong in the deep pit manure storage area. Dead birds need to be collected, appropriately stored, and handled according to a safe plan. If left in the droppings storage area, they contribute to the odor, attract rodents, and in general contribute to an ever-worsening condition.

Belt systems for frequent manure removal. Broiler housing units with an under-cage belt system for manure removal and broiler harvesting have been developed. The belt, which can be operated at any interval desired, takes the broiler droppings to one end of the house for emptying into an auger that conveys waste out of the building to spreaders or transport

Figure 12.6.   Endless belts have been used beneath poultry cages to achieve frequent droppings removal for storage or treatment outside the building.

vehicles. A typical example is shown in Figure 12.6. This belt system provides easy and frequent removal of waste to reduce odors and dust generation. For broiler harvesting, the cage floors open and the broilers are dropped onto the belt for transport to one location where they can be transferred to coops. Research is being conducted to see whether the same system can be adapted to handle turkeys.

Such a belt system also has been used in Korea. This belt has a convex surface so that liquids drain to a side trough for gravity flow to a treatment system. The solids are conveyed to one end of the house and emptied to a side auger for removal from the house. In this system, the removed solids are composted directly without any supplemental material being added. This system also results in reduced odors within the building and a dryer waste because the feces and urine are collected separately and the solids removed from the building frequently.

Other dry manure handling systems. Various other schemes for handling manure as a solid have been proposed but none has received widespread application, especially in the larger operations. There have been some experimental facilities in which swine are raised on a deep litter system. It is similar to the poultry litter system, except the litter layer is typically 3 to 5 feet in depth. This approach is discussed in greater detail in Chapter 9. There is an increasing number of confinement operations in which the liquid manure is passed over a solid–liquid separation, creating a solid manure stream. This material, essentially washed manure solids, can be subjected to further treatment, composting, or be hauled to cropland either before or after storage. There is seldom an odor issue when handling these washed solids. If there is, they can be covered or treated with an odor absorbent such as natural zeolite.

Application of Manure to Cropland

Land application is the final step in most livestock and poultry handling schemes. This step is a frequent cause of odor complaints because it creates a large surface area from which volatile compounds can escape. The most frequent response to this concern is to modify the application procedure to accommodate neighbors. Frequently, complaints can be avoided by selecting a time to haul manure when the wind blows odors away from sensitive areas. Application of manure during early morning when there are better drying conditions also helps. Obviously, it is thoughtful to avoid spreading manure immediately prior to a holiday or weekend when you might expect neighbors to be planning outdoor activities. Prompt incorporation of the spread manure into cropland also helps avoid odors. Soil injection is the most effective way to reduce odor during land application and is also helpful in reducing ammonia loss.

Application of lagoon contents is a necessity to maintain a stable chemical balance in the lagoon system with respect to salinity and ammonia as well as to maintain an appropriate water level. Planned removal of lagoon contents is part of the overall management strategy. One method, irrigation, is particularly prone to odor release and complaints triggered by visibility as well as odor detection. Irrigation equipment is helpful in reducing the time needed for application but requires considerable management attention to avoid creating problems. Low-pressure sprinkler systems with the emitters placed as close to the ground as possible create less water–air contact than high-pressure systems spraying upward. Irrigation from lightly loaded, well-managed, anaerobic lagoons and from second-stage or tertiary lagoons provides effluent of higher quality and less odor.

Feed Additives to Reduce Odor

There have been several attempts to reduce manure handling odors by altering the ration being fed or by the addition of a specific odor-reducing material, such as sagebrush, mint oil, or a sarsaponin extract of the yucca plant. Although the data from these materials are not conclusive, they suggest that it is possible to alter the odor of fresh manure; however, this change does not persist when manure undergoes anaerobic decomposition.

A related approach to manure odor control is to alter the feeding regime to achieve enhanced nutrient use. This approach has a logical attractiveness in that if the amount of manure could be reduced, there would be a reduced potential for the formation of odorous compounds. The use of synthetic amino acids has been reported to result in reduced nitrogen in livestock feces and urine. Currently, research is still in a preliminary stage and does not offer a clear possibility to the livestock or poultry producer as a technique to reduce potential odor problems.

Even less developed at this time is the possibility of manipulating the microbiological populations in the digestive tract to enhance animal performance and minimize nutrient excretion. This approach has potential triple benefits in reducing feed costs, manure handling, and the need for antibiotics to control disease organisms.

Air Treatment Alternatives

Various means to reduce the odor of air have been used in industries with long-term odor concerns such as the rendering industry, paper pulp processors, and fish processors. Absorption and adsorption processes have the potential to remove many of the chemical compounds identified in the headspace over manure. Activated carbon adsorption is widely used in these industries but has obvious limitations when considered for livestock enterprises because of the large volume of air typically handled.

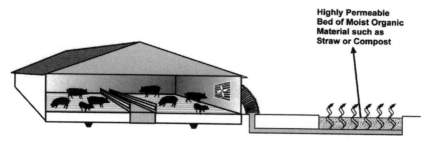

Figure 12.7.   Biofilter as might be used to decrease the odor being discharged from a confinement livestock facility.

European researchers have, however, used water-based scrubbing units to capture ammonia from the air within a swine building. In this system, air is blown upward through a bed packed with inert particulate material. Water is recycled countercurrently down through the packing so the air contacts a large water surface. The pH of the water is controlled to maintain a slightly acidic pH. The water accumulates high nitrogen content, indicating success and observers reported the building environment improved.

An alternative to this technology that has been practiced by rendering as well as a few livestock and poultry operations has been to pass odorous air through a shallow absorption bed. In this process, a unit similar to a septic tank absorption bed is constructed into which the exhaust air is discharged. By having the bed relatively large, the headloss was reasonable and by passing the air upward through the soil layer, odor removal is achieved. An alternative application of this approach is the use of straw, compost, or other highly permeable material as a biofilter to provide surface area for odorant absorption and aerobic conditions for biological oxidation. Figure 12.7 is a schematic view of a typical biofilter.

There are a variety of materials available for purchase that can be sprayed into the air downwind of a livestock or poultry operation that are designed to mask the malodor by a more pleasant odor. Citrus and floral scents are particularly popular. In general, these materials have not gained widespread acceptance but some operators consider them an effective way to achieve a short-term solution to an urgent problem. Materials also are available that promise to increase the rate of manure decomposition in a storage tank or anaerobic lagoon. These materials may claim to contain enzymes, special organisms, or other materials that result in a less odorous pit or lagoon. There are other products available that are designed to be fed to the animal and promise to reduce the odor from a lagoon. No one product has been identified as useful for the full

Figure 12.8.   A windbreak wall installed just beyond a ventilation fan can add additional turbulence causing the odorant to be diluted. Windbreak walls can be expected to be most beneficial when wind turbulence is least such as during a temperature inversion.

range of odor problems and some have been identified as ineffective even in the most controlled of conditions.

Some odor complaint reduction has been attributed to the construction of windbreak walls in the direct discharge of ventilation fans. These vertical walls cause additional turbulence, thereby dilution of the ventilation air. A typical windbreak wall is shown in Figure 12.8. Little data exist regarding the effectiveness of windbreak walls.

Designing for Odor Control

Designing for odor control is more difficult than for water quality control because of the difficulties inherent in measuring the odor control success of the facility. If complaints are used as the basis for assessing the degree to which the design is successful, success will be highly dependent upon site selection and the degree to which the livestock or poultry operation is seen as making a positive contribution to the community by the landowners immediately impacted by the odors. It has been observed that the degree to which a livestock operation is judged to have an unacceptable odor is related to the overall nature of the facility. Some of the

Table 12.7   Factors unrelated to livestock or poultry production facility design that are likely to produce more or fewer odor-related complaints

Likely to produce more odor complaints	Likely to produce fewer odor complaints
Proposed facility is several times larger than any existing similar facility in the area	Proposed facility is similar in size to those already existing in the area
Proposed facility owned by investors living outside the community	Proposed facility owned by established local livestock or poultry producers who are highly regarded within the community
Management personnel moved into the area from some other region of the country	Management personnel are local people who are highly regarded within the community
Operating personnel are from outside the region and they and their children do not speak English in the home	Operating personnel are long-term residents of the community
Proposed facility is perceived as making no or minimal contribution to the local community	Proposed facility is perceived as making a positive contribution to the community by the support of local activities
Proposed facility placed under construction before local people were aware of the plan	Proposed facility was constructed after extended local involvement in the planning process

important factors related to perceived odor nuisance that are largely unrelated to the design of the facility are listed in Table 12.7.

In addition to the factors listed in Table 12.7, there are design decisions related to overall odor level. Dirty animals and manure-covered pen floors contribute to elevated odors. Therefore, building designs that maintain clean floors, clean animals, and a low level of dust are less odorous. In addition, selection of the waste handling process has a profound impact on the odor level. For example, an uncovered manure storage tank allows the escape of more odorous gases than a similarly sized tank with a cover; an anaerobic lagoon emits less odor than an uncovered manure storage pit of the same surface area. Unfortunately, the state of the art is not sufficiently developed to be able to precisely compare the odor release of a 1-acre anaerobic lagoon with a 0.2-acre manure storage basin. This type of comparison is further complicated by the varying impact of temperature and season. Table 12.8 lists a number of odor comparisons that may be helpful in considering different alternatives for the handling and treatment of manure from a specific number of animals. The table was prepared for 1,000 finishing pigs to provide a solid basis of comparison, however, similar conclusions can be drawn for other species or numbers of animals.

Table 12.8   Comparison of the odor complaint possibilities of several waste management and treatment alternatives to collect, manage, treat and dispose of the manure from a 1,000-head swine finishing facility

More odorous option	Less odorous option
Building is poorly ventilated resulting in uncomfortable animals.	Well-ventilated building in which the animals have no need to roll in manure for cooling
Solid floor building equipped with a scraper system to remove manure	Totally or partial slotted floor building with an animal density that promotes rapid separation of manure from the animals
Manure flushed from the building with effluent from a heavily loaded anaerobic lagoon or manure settling basin	Manure flushed from the building with second-stage anaerobic lagoon effluent, aerated lagoon effluent, or fresh water
Heavily loaded anaerobic lagoon	Lightly loaded anaerobic lagoon
Open-top manure storage tank	Covered manure storage tank
Anaerobic lagoon	Aerated lagoon
Anaerobic lagoon	Enclosed anaerobic digester
Liquid manure or lagoon contents distributed to crop or pastureland with a high-pressure, small-droplet irrigation system	Liquid manure or lagoon contents distributed with a low-pressure irrigation system with minimal air contact
Manure or lagoon contents applied to the soil surface	Manure or lagoon contents injected into the soil or incorporated immediately after application

It is feasible from an engineering perspective to design, construct, and operate a confinement livestock or poultry facility to achieve odor control to whatever degree is judged suitable. Building design and waste management alternatives exist that control odors. The situation is made more complicated, however, because there is not an easily measured and universally accepted parameter to evaluate odor control. There is no agreement as to the frequency, intensity, or duration at which an odor is acceptable. An examination of the options listed in Table 12.8 reveals manure handling systems that produce less odor tend to be more expensive to construct and operate than those from which more odors are released. Thus, the designer is uncertain that enough odor control has been incorporated into the total production and waste management system and is uncertain that odor control costs are necessary for the success of the enterprise. Thus, most current facilities have expended little that can be directly attributed to odor control.

Index